THE TRIUMPHS AND TRAGEDIES OF THE ORPHAN CAR COMPANIES

Alan Naldrett

New Naldrett Press

ISBN: 9781076047571

Introduction

Many of the car "firsts," such as air-conditioning and locking doors, were initiated by orphan car companies that are gone and forgotten today. For instance, the Marmon Wasp had the first rear-view mirror, and the first push-button door lock was in the Franklin in 1917. Both marques are unknown today by the general public but were leading innovators in their day.

What is an "orphan" car company? Originally, it was a car company that has outlived its original company—which is pretty much most cars nowadays. Today it's considered to be any auto that is connected to a marque that has discontinued production completely.

With hundreds of car companies, there were a lot of chances for a car company to introduce something that would soon spread to most of the companies, such as the steering wheel, introduced by and further popularized by the top-selling Jeffrey Company's Rambler in 1904.

While many companies distinguished themselves in some ways, many companies didn't survive long enough to join the list of car firsts but have interesting stories. It would be impossible to chronicle all the car companies of the United States. However, I've tried to include many of the orphan car companies that had manufactured and assembled cars that were notable but have been rarely written about.

Car historians often divide the topic into different "eras," including the Veteran's Era, Brass Era, Classic Era, and Modern Era and I've followed this in chronological classifications for my chapters.

The Veteran's Era contained many of the earliest American auto pioneers, including the Duryea Brothers and Harry Winton. Many of the "firsts" of the auto industry occurred during this era.

The Veteran's Era led to the Brass Era, when many of the significant innovations occurred. The Classic Era and Modern Era also had many interesting stories of their own! If you have any relevant stories about the independent auto companies, e-mail me!

While learning about the auto companies of the past, meet the mostly forgotten auto pioneers who lived at the dawn of the auto age. They headed car companies that were brief flashes in the stratosphere, while some went down in flames almost from the start. Very few survived more than three years.

But in an age where there were so many auto companies it's difficult to count them all, there are some unheralded companies who stand out. Did you know the Otis Elevator Co. made autos before skyscrapers overwhelmed them with elevator orders? How about the Reeves Pulley Company who went from making industrial pulleys to motor cars, and then back to pulleys?

These and many more tales of the orphan auto companies are within these pages! I hope you enjoy them all at least half as much as I enjoyed learning about them!

-Alan Naldrett

Introduction

Table of Contents

Part One—The Veteran's Era—1759-1894

Chapter One—The Steam Car Makes Its First Appearance in France in 1759

Chapter Two—Early American Steam Cars Include the Thing and Lynn, MA

Chapter Three—The Benz From Germany is the First Successful Gasoline-Powered Vehicle in 1888

Chapter Four—The Duryea Brothers Market the First Successful American Automobile in 1893

Chapter Five—Alexander Winton Follows Close Behind the Duryeas With His Successful Auto in 1894

Chapter Six—William Jennings Bryan Becomes the First Presidential Candidate to Campaign in a Car (a Mueller) in 1896

Part Two—The Brass Era—1895-1919

Chapter Seven—The Oldsmobile Has the First Speedometer

Chapter Eight—Tincher has the First Power Brakes in 1903

Chapter Nine—Who Had the First Steering Wheel?

Chapter Ten—Locomobile Has the First 4-Speed Transmission in 1904

Chapter Eleven—The Twyford Stanhope is the First Gasoline Auto with 4-Wheel Drive in 1904

Chapter Twelve—The Marmon Wasp Has the First Rear-View Mirror in 1910

Chapter Thirteen—Cadillac Has the First Self-Starter in 1912

Chapter Fourteen—Maxwell Has the First Adjustable Driver's Seat in 1914

Chapter Fifteen—The Scripps-Booth Auto Has the First Horn in the Middle of the Steering Wheel in 1915

Chapter Sixteen—The Willys-Knight Auto Had the First Mechanical Windshield Wipers in 1916

Chapter Seventeen—The Franklin Had the First Pushbutton Lock in 1917

Part Three—The Vintage Era-1920 to 1930

Chapter Eighteen—The Hudson Motor Company had the First Adjustable Seats in 1921

Chapter Nineteen—The Duesenberg Had the First Hydraulic 4-Wheel System in 1921

Chapter Twenty—The First Car Radio Was in the Chevrolet in 1922

Chapter Twenty-One—The First Back-Up Lights and Use of Molybdenum Steel on an Appear on the Wills Ste. Claire in 1922

Chapter Twenty-Two—Studebaker Has the First Windshield Defroster in 1928 and the First Windshield Washer in 1937

Chapter Twenty-Three—The First Power Brakes Appear on the Pierce-Arrow in 1928

Chapter Twenty-Four—The First Mass-Produced Front-Wheel Drive Auto is the Cord in 1929

Part Four—The Classic Era-1931 to 1948

Chapter Twenty-Five—The First All-Mechanical Automatic Transmission is Developed by REO in 1932

Chapter Twenty-Six—Packard Introduces Air-Conditioning in 1939, Power Windows in 1941, and Power Locks in 1954

Chapter Twenty-Seven—Nash Had the First Seatbelt in a Mass-Produced Car in 1948

Chapter Twenty-Eight—Crosley Had the First Mass-Produced Disc Brake in 1949

Part Five—The Modern Era—1949 to Present

Chapter Twenty-Nine—The Plymouth Valiant Had the First Alternator in 1960

Part Six—Orphan Car Companies by Region

Chapter Thirty—The Orphans of the South

Chapter Thirty-One—The Orphans of the Far West

Chapter Thirty-Two—The Orphan of the Rocky Mountain State

Chapter Thirty-Three—The Orphans of the New England States

Chapter Thirty-Four—The Orphans of the Midwest States

Chapter Thirty-Five—The Orphans of the Atlantic States

Chapter Thirty-Six—The Orphans of the West

Chapter Thirty-Seven—The Car Companies of Canada

Part Seven—Automotive Topics

Chapter Thirty-Eight—The History of 4-Wheel Drive Vehicles

Chapter Thirty-Nine—Everyone and Their Brother Had an Auto Company

Chapter Forty—The Previous Professions of the Early Automakers

Chapter Forty-One—The Most Popular Auto Company Names

Chapter Forty-Two—More Auto Company Firsts

Chapter Forty-Three—Sales Charts for U.S. Autos 1897 to 1989

Afterword

Part One—

The Veteran's Era—

1759-1894

Chapter One—The Steam Car Makes Its First Appearance in France in 1759

The first widely-built engines were steam engines, and the first autos were powered by steam. The first recorded vehicle, or the "first working, self-propelled, land-based mechanical vehicle," which could go about 4 miles per hour, was constructed by

Frenchman Nicolas-Joseph Cugnot in 1759. It had to have its boiler relit every 15 minutes or so.

Cugnot's steam vehicle (which he called a "fardier," is pictured and was preserved in the Conservatoire National des Arts et Métiers Museum in Paris. It can now be seen in the Musée des Arts et Métiers in Paris.

The first auto accident was said to be in Cugnot's steam car. It was reported that Cugnot knocked down a brick wall in 1771 and was arrested and fined for dangerous driving.

The first auto accident in 1871 on the streets of Paris

The vehicle was deemed impractical but was deemed important enough for King Louis the 15th to grant Cugnot a lifetime pension. That is, until the French Revolution, when the pension was suspended. It was restored again by Napoleon, as Cugnot lived on to be 79.

Inventions of steam-powered engines were credited to Taqi al-Din in Egypt in 1551, Giovanni Branca in Italy in 1629, while Jerónimo de Ayanz y Beaumont of Navarre (Spain) received patents in 1606 for 50 steam- powered inventions

Chapter Two—Early American Steam Cars Include the Thing from Memphis, Michigan, and the Simonds Steamer from Lynn, MA

Across the pond, in the U.S., inventors were independently building their own steam vehicles. In Memphis, Michigan, there is a Michigan historical sign in the town of Memphis, Michigan right at 35412 Bordman, on the county line dividing St. Clair and

Macomb Counties. It commemorates "The Thing," a steam auto built by Tom Clegg and his father in the 1880s. The Thing (also known as the *Cleggmobile*, after its inventors) has been called the first self-propelled vehicle in Michigan (and maybe the nation). Called "The Thing" by the Cleggs, it was also called "the contraption."

John Clegg was born in England and moved to Toronto, Ontario by 1863, when his son Thomas was born there. The family moved to Detroit in the same year and then in 1870 the family moved to Memphis, Michigan.

John formed the *John Clegg & Son* machine shop with his son Thomas and built a steam tractor in the early 1880s. The shop provided a place where the Cleggs could work on steam engines and self-propelled vehicles and so after the steam tractor, Thomas nagged his father to help him build a steam car. Thomas finally relented and, in the winter of 1884, they worked on the vehicle. It was finished by the spring of 1885 and in June was taken out for the first of over 30 test drives covering around 500 miles.

It was a one-cylinder, steam-driven vehicle which could attain a top speed of about 12 miles per hour. The tubular boiler was in the rear of the vehicle and used coal for fuel. The rear wheels were 5 feet and 8 inches tall and It had room to seat the driver, a "stoker," and two passengers. Its longest trip was to the neighboring town of Emmett. It was on this trip that the engine "blew up." The Cleggs dismantled the vehicle and sold the engine to the local creamery.

The Thing sat in front of the Cleggs house for several years and sod grew up around its wheels. Some kids pushed it down the hill and in time the rest of the vehicle was reportedly buried somewhere on the site of the machine shop.

In 1886 Thomas constructed another car, a brass one weighing less than the Thing, about 2,800 pounds. In 1904 he built a gasoline vehicle but never built any more after his father's death.

Henry Ford and Ransom Olds were acquainted with the Cleggs and consulted with them about motorized vehicles. Henry Ford offered to buy the Clegg machine shop to display in Greenfield Village, but in 1936, shortly before the offer, Thomas had "pulled down" the building.

On the 95th Anniversary of the Thing in 1975, the townspeople of Memphis threw a celebration and the historical plaque honoring the Thing was unveiled. A reward of $100 was offered for a picture of the Thing, but nobody had one. Pictured is the replica built for the Memphis museum.

The Cleggmobile, aka "The Thing" replica

Clarence Simonds is on the tiller in this picture of the Simonds Steamer

In 1863, the *Simonds Steamer* drove the streets of Lynn, Massachusetts. The invention of Clarence Simonds, an employee of the Lynn Gas and Electric Company, it had a 2-cylinder, vertical engine. It used naphtha as fuel and could attain a full head of steam in five minutes. It had a "porcupine-type" boiler. Once going, the auto could attain the speed of 10 miles per hour. There were two pumps—"one to feed the boiler, the other to blow the air blast through the naptha."

According to early magazine *The Horseless Age*, "The exhaust steam passes through a feed water heater is then delivered to the naptha flame, where its presence stifles the noise."

Originally Simonds had built the vehicle for his own use (he obtained special permission from the Town Council to run it on the streets of Lynn). But it soon became famous around town and he sold it to a consortium of townspeople for what he described as a "round sum."

It's believed that Simonds never built any more vehicles but consulted with the Stanley twins (makers of the *Stanley Steamer*) on steam cars.

Chapter Three—The Benz From Germany is the World's First Successful Gasoline-Powered Vehicle in 1888

If one was to ask, "Who invented the automobile?" the answer would probably have to be Karl Benz, an engineer from southwestern Germany, and Gottlieb Daimler, who created the first working gasoline-powered motor vehicle in 1885. With three wheels with bicycle tires, a 4-stroke engine, and an electric ignition, the vehicle was patented in 1886 as the *Benz Patent-Motorwagen.*

The *Benz Patent-Motorwagen of 1885*

Chapter Four—The Duryea Brothers Market the First Successful American Automobile in 1893

The Duryea brothers, Charles, and chiefly, Frank, constructed America's first working gasoline-powered car in Springfield, Massachusetts, in 1893. Named the *Duryea,* it had a single-cylinder, 4 horsepower engine, with friction drive, which was located under the seat of the buggy-like body.

Charles and Frank Duryea in their gas vehicle, the Duryea, in 1895.

Frank Duryea drove a *Duryea* in the first *Chicago Times-Herald* auto race of 1895 and won. Charles formed the **Duryea Motor Wagon Company** in 1895 to manufacture cars to sell to the public. The Duryea went on to be licensed to be made in Coventry, England and Waterloo, Iowa and continued in various locations until 1917.

The car was started, as most of the early gas-powered cars were, by a crank. The procedure one went through to start the early crank-starting autos was one of the main reasons many people wouldn't buy a gasoline-powered auto. To start a "crank" auto you had to:

1. Turn on the carburetor tap

2. Switch on the accumulator

3. Engage the brake

4. Disengage the clutch

5. Put the gear shift into neutral

6. Open the throttle

7. Retard the ignition

8. Tickle the carburetor float

9. Crank the car until the engine turns over

Lots of people were seriously injured from the car crank recoiling, until Henry Leland and Charles Kettering's self-starting auto ignition and the subsequent formation of Delco in 1910. Byron Carter, of **Cartercar** fame, died from gangrene of the jaw after suffering the kickback of a crank. This is also where the term "cranky" came from!

Chapter Five—Alexander Winton Follows Close Behind the Duryeas With His Successful Auto in 1894

No history of early autos would be complete without Alexander Winton. Winton drove his first auto down the streets of Cleveland in 1894. He moved from Scotland in 1878 at the age of 19 and worked as an engineer on an ocean steamship. In 1891, he started a bicycle shop with his brother-in-law (bicycling was a big thing in the 1890s). While working in the bicycle shop, Winton devoured everything he could read about "horseless carriages." Scientific American and The Horseless Age magazines were the best sources for information about self-propelled vehicles. Winton not only successfully built one of America's working autos, he successfully made and sold many more. He also earned more than 100 patents having to do with automobiles.

As intelligent as he was, Winton was vain enough to hate being contradicted. A prime example is when James Packard bought a car from Winton and had it break down several times while driving it home to Warren, Ohio. He fixed it himself and when he next saw Winton, he told him that he had thought of a few improvements he could make on the *Winton* car. Winton said to Packard, "If you're so smart, why don't you start your OWN auto company?" And of course, that's what James Packard did, started the **Packard Motor Company** with his brother Louis.

Chapter Six—William Jennings Bryan Becomes the First Presidential Candidate to Campaign in a Car (a Mueller) in 1896

Automobiles, or motorcars, were still very new when the Democratic candidate William Jennings Bryan was campaigning for President. He appeared in Decatur, Illinois in a rented *Mueller,* an auto made in Decatur by Hieronymus Mueller and his six sons. One of the earliest American autos, it successfully made a test run from Decatur to Bloomington, Indiana in 1896. The Mueller family talked about manufacturing autos, but as far as is known, only four *Mueller*s were ever made. William Jennings Bryan went on to be the Democratic candidate two more times, but he never won.

William Jennings Bryan

H. Mueller with his automobile, the Mueller

Part Two—

The Brass Era—

1895-1919

Chapter Seven—The Oldsmobile Has the First Speedometer in 1903

    If any person deserves to lead off the Brass Era section, it would be Ransom Olds. Ransom was a boy genius and dabbled in inventing electric and steam cars before designing the "Curved-Dash Olds," which an employee drove out of the Detroit factory before it burned to the ground. As the other prototypes were destroyed, it was decided to build the Curved-Dash, once the company relocated to Lansing, due to being offered incentives by the municipal government of the area.

As if the first auto that was affordable to the middle class wasn't enough, the Olds is also credited with introducing the first speedometer in a motor vehicle. First invented and patented by German engineer Otto Schulze in 1902, the Curved-Dash Olds was the first to use it. By 1910, most auto marques had replaced their tillers with steering wheels.

Chapter Eight—Tincher Has the First Power Brakes in 1903

Thomas Tincher developed the Tincher and introduced it at the New York Auto Show, where it was a big hit due to its air pump, which operated the first power brakes in an auto. The air pump could also be used to put air in the tires and to toot the car's horn!

1907 Tincher

The Tincher had 90-horsepower, a large, 126-inch wheelbase, with a 4-speed transmission and double-chain drives. After releasing 4-cylinder cars in 1903-1907, while they were in Chicago, they were a pioneer in introducing a 6-cylinder vehicle in 1908, after the **Tincher Motor Car Company** was moved to South Bend, Indiana, because the Studebaker brothers invested in the vehicle.

The Tincher was one of the most expensive vehicles of the day, starting at $5,000 and going up from there. In 1904 they had a racecar selling for $12,000. Unfortuneatly, they were only able to build six or seven cars a year. With this low level of production, they were unable to stay profitable and liquidated in 1909. Thomas Tincher declared personal bankruptcy and later worked for the **Economy Motor Buggy Company**.

Chapter Nine: The First Steering Wheel

For the first years of automobiles, they were steered by using a "tiller," a long thin rod, still used by ships and boats. The nautical definition is "a bar or lever fitted to the head of a rudder, for turning the rudder in steering." For steering an automobile, it was felt that the tiller was inadequate.

The first recorded steering wheel is the one used by Alfred Vacheron on his custom four-horsepower *Panhard* he used to participate in the Paris-Rouen race in 1894. Other racecar drivers followed suit.

The first car company to formally introduce a steering wheel in one of their vehicles is the **Packard Motor Company**, who had one installed on their Model C in 1900. Probably because this car was not widely distributed it is often overlooked as having the first steering wheel for a motorcar. (They only made eight of the Packard Model A).

Packard Model C, the "Ohio," had the first steering wheel

Often more generally accepted as the first car to introduce the steering wheel to the masses is the 1904 Rambler. The man who started the **Jeffrey Company**, in Kenosha, Wisconsin, was Thomas B. Jeffrey. Jeffrey was at first a bicycle manufacturer and a partner in Gormully & Jeffery. He sold his stake in the company to his partner and used the capital to get started in the auto business, building his first auto in 1897. By 1902 he had purchased a large bicycle factory in Kenosha to produce *Ramblers,* named for a popular bicycle of his previous company, Gormully & Jeffrey.

Thomas B. Jeffrey   Thomas B. Jeffery logo

Jeffrey's son Charles did the designing for the **Jeffrey Company** and added a steering wheel in 1902. He wanted to put the engine under the hood for the first *Ramblers*, instead of under the seat like most cars of the day. Thomas talked Charles out of these two improvements on their first cars in 1902 because he deemed them too radical.

By 1904 however, Charles persisted and included the steering wheel in the 1904 model and became the first company to do so. (By this time, the front-mounted engine had been done by *Locomobile)*. Jeffrey was the second auto company to use an assembly line (**Oldsmobile** was the first). They built and sold over 5,000 *Ramblers* in their first years.

Thomas B. Jeffrey died in Pompeii, Italy in 1910 and his son Charles took over as company president. In 1915 he changed the name of the auto from *Rambler* to *Jeffrey*, to honor his father.

When Charles Jeffrey survived the sinking of the Lusitania, he decided to retire in 1916. He sold the company to former GM head Charles T. Nash, who renamed it **Nash Motors** after himself. Nash survived a long time, including being part of **American Motor Company** (**AMC**), later absorbed by Chrysler.

After it was introduced in the Rambler, almost all of the auto companies wasted no time in also replacing the awkward tiller with the more manageable steering wheel.

Curiously enough, the 1901 Packard Model C had a steering wheel and actually released the model, a roadster with a 183 cubic-inch single cylinder engine that generated 12 horsepower on a 76-inch wheelbase. Limited distribution of this car has caused the Jeffrey, because of its widespread popularity, to be the one that most often gets the credit for the first steering wheel.

Chapter Ten—Locomobile Has the First 4-Speed Transmission in 1904

The Locomobile wasn't called that because you had to be "loco," or Spanish for crazy, to buy one. It was because the name referred to "locomotion," the ability of the vehicle to move. Regardless, the company was founded in 1899, a pioneering car company that started in Bridgeport, Massachusetts and moved its manufacturing to Bridgeport, Connecticut, in 1900.

Locomobile produced a small steam car with a design purchased from the Stanley twin brothers, of Stanley Steamer fame, until 1903. They then switched to gasoline engines, which they continued to manufacture until their demise in 1929. They had been the luxury brand of conglomerate Durant Motors since 1922.

In 1904, the company released the first 4-speed transmission. The popularity of it soon led other car companies to follow suit.

1909
Locomobile
Model "30"
Baby Tonneau

Chapter Eleven—The Twyford Stanhope Was the First Gasoline Auto with 4-Wheel Drive in 1904

Robert Twyford built his first car in 1899 and prepared to market his *Twyford Stanhope* in 1901. Through a complicated maze of shafts, axles, friction clutches, and bevel gears. Another set of bevel gears provided the car with an early form of power steering! In 1901 and 1902 the Twyford Stanhope (named for the Stanhope family) was manufactured. Twyford knew that 4-wheel drive would provide enough power for cars to navigate snowy, muddy, and rocky road.

Twyford had trouble finding further backing for his auto. At first located in Pittsburgh, he finally found enough capital to build a prototype in 1904 from Brookville, Pennsylvania, who welcomed the **Twyford Motor Company**. A brick, two-story factory was built next to a furniture factory in Brookville, and the first two cars were built there.

These Twyfords were exhibited with success at the Buffalo Auto Show in 1905 but sales were not outstanding. Two-cylinder cars were produced from 1905 to 1907 but the company liquidated after that.

Robert Twyford took his 4-wheel drive patents to the **Commercial Motor Car Company** in Houston, Texas in 1911, but in 1912 this company also went into receivership. Twyford didn't try anymore to produce his 4-wheel drive cars, but he has a name in auto history as the first proponent of 4-wheel drive.

1904 Twyford, the car with the first 4-wheel drive.

Chapter Twelve—The Marmon Wasp Had the First Rear-View Mirror in 1911

Howard Marmon was a college-educated (UC Berkeley), talented mechanical engineer when he joined the family firm, Nordyke and Marmon, in the mid-1890s. By 1902 he was the chief engineer of the Indianapolis firm, which made flour-milling machinery under the brand name of Marmon.

By 1902, when he was 23, Howard Marmon became the chief engineer and also built his first auto. In 1905 he started building autos to sell and made five. *Marmon*s continued to have low production but stayed extant while they slowly grew, until in 1923 the **Marmon Motor Car Company** was producing 4,500 autos per year, at a premium price, often $5,000 or more.

When the *Marmon Wasp*, a bright yellow model, won the first Indianapolis 500 in 1911, it sported the first rear-view mirror in a car. Racecar driver Ray Harroan rigged up a mirror so he could see the cars to the rear of him, in lieu of having a passenger beside the driver to keep an eye out. Harroun hooked up a 3-inch-by-8-inch mirror mounted above his dashboard so that he could see behind him. As the only one-passenger car out of 40 racecars, this was surely an advantage as Harroun drove a one-man, streamlined auto and went on to win. Meanwhile, the rear-view mirror caught on and soon all racecars, and then all cars, sported a rear-view mirror.

Ray Harroun, winner of the first Indy 500 in 1911, in his Marmon Wasp. No windshield but notice the rear-view mirror in the upper right of the photo.

Just a month after **Cadillac**, the *Marmon* introduced the second 16-cylinder engine in a car in 1930. **Marmon**s continued as a high-priced, high-quality auto until Depression-era 1933, when the brand disappeared, like many other luxury-car brands.

Chapter Thirteen—Cadillac Has the First Electric-Starter in 1912

Henry Leland took over Ford's second car company and built the *Cadillac*. He also worked with Charles Kettering to develop the electric starter to replace the crank. The crank was the biggest impediment to the growth of the gasoline-powered car. It was responsible for countless injuries, as the crank often had a kickback and caused broken arms and worse. (This was the origin of the word "cranky.") The first starter appeared on the 1912 *Cadillac Touring Edition.*

Cadillac went on, long after Leland left to start **Lincoln Motor Car Company**, to have many more innovations original to the car industry. These included the first 16-cylinder engine, the first fully enclosed auto, and automatic climate control.

Chapter Fourteen—Maxwell Had the First Adjustable Driver's Seat in 1914

Jonathan Maxwell was one of the many automakers who got their start with the Oldsmobile organization. He formed the Maxwell Motor Company in 1904, with the Briscoe brothers, Frank and Benjamin, which went on to have a healthy run. When it joined Benjamin Briscoe's **United States Motor Company**, it became the only surviving brand of the conglomerate. After that, it was restructured by Walter Chrysler, and the company became the **Chrysler Motor Company**.

In 1914, in the quest to improve the driver's comfort, the **Maxwell Motor Car Company** developed a driver's seat in the *Maxwell 25* that could be adjusted by three inches to the driver's desire.

Chapter Fifteen—The Scripps-Booth Auto Had the First Horn in the Middle of the Steering Wheel in 1915

The **Scripps-Booth Auto Company** could be said to be the plaything of James Scripps-Booth, heir to the Scripps newspaper fortune. While the cars were of unique design, they were never profitable, yet Billy Durant, founder of General Motors, chose to purchase the brand and never produce any more. Whether he just wanted to eliminate a competitor or wanted to use the factory space isn't known.

Before the company was purchased, in 1914 they produced an auto that took the horn from its familiar place to the left of the driver, mounted in the window-frame of the driver's side of the car. They instead put it right in the middle of the steering wheel, which started a trend that everybody today follows.

Scripps-Booth radiator badge

Chapter Sixteen—The Willys-Knight Auto Had the First Mechanical Windshield Wipers in 1916

The **Willys-Knight Motor Company** started in 1908 when John Willys purchased the failing **Overland Automotive Division** of the Standard Wheel Company. Willys built the company up until it had the second-largest amount of auto sales from 1912-1918 (**Ford Motor Company** was number one). Willys purchased a license to use Charles Knight's sleeve-valve engine and started manufacturing the *Willys Knight*. (John Willys also purchased the **F.B. Stearns Company** of Cleveland and used the additional factory space to begin manufacturing the *Stearns-Knight* luxury car).

In 1916 the **Willys-Knight Motor Company** made auto history when they used Mary Anderson's design for what she called, a "window cleaning device." These became the auto world's first official windshield wipers.

The **Willys-Knight Motor Company** went on to produce more autos, such as the *Willys Six* and then received a contract from the U.S. Government to produce *Jeeps* for the military. After the war the **Willys-Knight Motor Company** stopped making cars and just made jeeps, until 1952 when they once more manufactured an auto, the *Willys Aero.*

In 1953 the **Willys-Knight Motor Company** was purchased by the **Kaiser Motor Company** and continued to make Jeeps. When **Kaiser** exited the auto business to concentrate on aluminum, they sold the auto division, including Jeep, to the **American Motors Company**.

Chapter Seventeen—The Franklin Had the First Pushbutton Lock in 1917

The *Franklin* auto was built in Syracuse, New York, starting in 1901, making their manufacturer **H. H. Franklin & Co**. one of the earliest car companies. Its founder, Herbert H. Franklin, started the first machine die-casting enterprise in the world in 1893 and in 1901 built an air-cooled engine with engineer John Wilkinson.

In 1902 the *Franklin* auto made its debut, powered by gasoline and manufactured by the **Franklin Automobile Company**. H.H. Franklin was the company president and Wilkinson the chief engineer. Wilkinson had earlier built two cars with air-cooled engines for the **New York Automobile Company** and was never paid by them, leaving him free to take his autos with innovative air-cooled engines elsewhere. Air-cooled engines were good because they did away with water pumps, radiators and radiator hoses and didn't freeze in the winter like water-cooled engines. Once Wilkinson gave a ride to Herbert Franklin in one of the autos, Franklin was hooked and decided to go into auto manufacturing.

The first Franklin took two months to build and had the first 4-cylinder engine in a car offered to the public. It was lightweight and had a top speed of 12 miles per hour. Because it only weighed 900 pounds and used elliptic springs, it was able to navigate the rough roads of the period. In 1902, thirteen *Franklin*s were sold, at $1,100 apiece.

The *Franklin Light Roadster* came out in 1902 and featured two forward speeds and reverse, as well as two brakes. It had a frame constructed of three-ply laminated wood from an ash tree. In 1904, L. L. Whitman drove a *Franklin* from New York to San Francisco in just under 33 days, besting the records previously set by *Packard* and *Winton* autos. Innovations included a carburetor that maintained a constant level, and 6- and 8-cylinder models in 1905.

Franklin Light Roadster

By 1915, *Franklins* were being built that could go 65 miles per hour and could achieve 32 miles to a gallon of gasoline. It was known as a luxury car, but less expensive than a *Cadillac* or *Packard*. The engines were all the overhead valve type.

As a security feature, the 1917 *Franklins* introduced a pushbutton lock, the first in the industry and widely-copied forever after. Throughout the 1920s there were about 8,000 Franklins built and sold each year, with 14,000 in 1929 the peak year.

The **Franklin Automobile Company** introduced a 100-horsepower engine in 1930, and in 1932 came out with a 12-cylinder engine. But due to the onset of the Great Depression, only 200 of the 12-cylinder *Franklins* were made. Sadly, in 1934 the Franklin Motor Company declared bankruptcy, ending a great automobile run.

The Franklin continuously had an overhead valve, air-cooled engine.

The 1934 Franklin Twelve had a 12-cylinder engine

Part Three

The Vintage Era

1920-1930

Chapter Eighteen—The Hudson Motor Car Company Had the First Adjustable Passenger Seats in 1921

The **Hudson Motor Car Company** started in a University of Michigan frat house when Roy Chapin and Howard Coffin met. As Chapin became car-crazy after his first ride in an *Oldsmobile* and quit college to work at the **Olds Motor Company**, he talked future **Hudson** employees Coffin and others into first, joining him at **Olds Motor Company**, and then forming a new company with him. It was named the **Hudson Motor Car Company** after financial angel, department store mogul Joseph L. Hudson, founder of J. L. Hudson Department Store.

Hudson was considered a medium-priced brand, above *Ford* and *Chevrolet*. In 1909 they started building autos in the old *Aerocar* factory at Mack Avenue and Beaufait Street in Detroit. They moved to a new factory at Jefferson and Connor Avenues in 1910. In 1913 they released the *Hudson Super Six*, which had a 6-cylinder engine. This gave it considerably more power than the mostly 4-cylinder cars then on the market.

The **Hudson Motor Car Company** introduced many innovations to the auto field and in 1921 were the first auto company to provide adjustable seats in an automobile.

With their fellow model the Essex, in 1929 the **Hudson Motor Company** was the third largest in the nation, behind **Ford** and **GM**, with over 300,000 autos sold. With the onset of the Great Depression, the **Hudson Motor Car Company** was one of the few auto companies to emerge in the 1940s. After helping the U.S. in World War II by retooling factories to build military equipment, **Hudson** returned to auto production in the late 1940s. In 1954 they merged with the **Nash-Kelvinator Corporation** to form the **American Motors Company (AMC)**. The last car to use the *Hudson* brand name was in 1957, and **AMC** lasted until 1987, when they were absorbed into the **Chrysler Motor Company**.

Chapter Nineteen—The Duesenberg Had the First Hydraulic 4-Wheel System in 1921

One of the most famous and desired luxury cars, especially in the United States, was the *Duesenberg*, a luxury car from Auburn, Indiana. It was founded in St. Paul, Minnesota by brothers August and Frederick Duesenberg in 1913. In 1916 they went to Elizabeth, New Jersey to build engines for the war effort.

In 1919 they moved to Indianapolis, Indiana, and formed the **Duesenberg Automobile and Motors Company, Inc.** Their first auto was called the *Model A* and had the *Duesenberg Straight-8 Engine*, an 8-cylinder engine that was very popular.

The first Hydraulic 4-Wheel System was contained in the *Duesenberg Model J,* one of the most desired cars of the Vintage Era. The *Model J* came out in 1921 and had a supercharged, 320 horsepower engine which could reach 104 miles per hour *in second gear* and was capable of speeds up to 140 miles per hour.

The company was purchased by E.L. Cord in 1926 and became part of the **Auburn Motor Company**, lasting until 1937.

Chapter Twenty—The First Car Radio Was in the Chevrolet in 1922

After Billy Durant was ousted from **General Motors** for overspending, he started a "comeback" by co-founding the **Chevrolet Motor Company** with racecar driver Louis Chevrolet. Using profits from **Chevrolet**, Durant brought controlling stock in **General Motors** and once again took over control of the corporation he founded. The **Chevrolet Motor Company** was folded into **GM**, and in 1922 came out with the first radio in a car.

The radio in the *Chevrolet* in 1922 was not very convenient. The speakers took up the whole backseat, the antennae took up the entire roof, and there were batteries that barely fit under the seat. Not to mention the fact that there were still very few radio stations in 1922, and the whole setup cost $200, a very hefty price in 1922.

Chevy car radio of 1922

By 1930, Motorola, led by Edward Lear, had invented a much more convenient car radio and radios became a standard feature in autos. By 1946, nine million cars had radios, and by 1963, 50 million cars.

At first, many of the populace was opposed to a radio in a car, for the same reason people oppose phones and texting while driving today: too distracting for the driver. In fact, the state of Massachusetts and the city of St. Louis, Missouri banned the use of car radios in 1930. Some of the reasons given were that they took driver's attention away from the road and could cause accidents, and music could lull a driver to sleep.

But while anti-radio laws were phased out, anti-texting laws have been enacted in 35 states and the District of Columbia.

Chapter Twenty-One—The First Back-Up Lights and Use of Molybdenum Steel on an Auto Was on the Wills Ste. Claire in 1922

The *Wills Sainte Claire* was the brainchild of the man who has been known as Henry Ford's "right-hand man" in the early days of auto making, C. Harold Wills. Recruited for both his artistic skills in drawing blueprints, and also his engineering skills, Wills was listed by the Detroit Business Directory of 1899 as being the Draftsman of the **Henry Ford Auto Company,** while Henry Ford was listed as the Superintendent.

Wills went on to help Ford in 1902 when Ford was building racecars during the days of his **Henry Ford Auto Company.** Finally, Wills joined Ford in his **Ford Motor Company** and designed Ford's early, successful vehicles, the *Models A. B, C, F. K, N, S, and T.* Wills also created the familiar Ford logo, the word Ford in Henry's own cursive writing. The logo is still used today. His orbital transmission of the *Model T* was widely copied, as was the Vanadium Steel the **Ford Motor Company** switched to.

Because his designing days were done once Ford decided the *Model T* wasn't going to ever be changed, Wills turned his skills to metallurgy. While still collecting the million dollars a year annual salary Ford was paying him, Wills went on to use his knowledge to synthesize newly-discovered molybdenum steel.

When Will took the molybdenum discoveries he had made at Ford to his new company in 1919, he had also purchased molybdenum mines. The car he built in Marysville, the *Wills Sainte Claire,* would be the first to use the new metal synthesis Wills created. When Wills stopped making the *Wills Sainte Claire,* he went to work for Chrysler and helped develop *Aeolite Steel.*

Chapter Twenty-Two—Studebaker Has the First Windshield Defroster in 1928 and the First Windshield Washer in 1937

The **Studebaker Motor Company** started off making carriages and buggies, and at one point were the largest buggy manufacturers in the world. When the youngest Studebaker brother Johnny went to the West Coast, he came back with all sorts of fanciful ideas about manufacturing motor vehicles since buggies would soon be passé, out-of-date.

The first Studebaker vehicle was an electric vehicle. Soon, the company switched to mass-producing gasoline-powered vehicles and became a leader in the auto field. Along the way, they absorbed the **EMF** and **Packard Motor Companies.**

Some of the innovations brought to the auto world by **Studebaker** were designed to help the driver have good visibility of the road. The first was the first windshield defroster in 1928, and windshield washers first appeared on a *Studebaker* in 1937.

Studebaker went the distance, not getting out of the auto business until their Ontario, Canada factory closed in 1963. Segments of the company that survived now make air-conditioners.

Chapter Twenty-Three—The First Power Brakes Appear on the Pierce-Arrow in 1928

In 1901 the **Pierce Arrow Motor Car Company** of Buffalo, New York, brought out its first successful car after manufacturing bird cages, ice boxes. In 1896 they started manufacturing bicycles and in 1900, they released their first auto, a steam vehicle, but it wasn't successful.

The 1901 gasoline-powered *Pierce Arrow* was engineered by David Ferguson and was had very strong sales—over 15,000 of the car had been sold by 1915. After many successful years, the *Pierce-Arrow* was the first auto company to have power brakes in their cars.

Pierce Arrow was considered one of the three P's of luxury brands. They were the *Pierce-Arrow,* the *Peerless* of Ohio, and the *Packard* of Michigan. A much-loved brand, the year of 1923 saw the **Pierce Arrow Motor Car Company** declared bankruptcy and placed into the receivership of the **Studebaker Corporation.** In 1933 a consortium of Buffalo businessmen brought the rights to produce the *Pierce Arrow,* which they did until 1938.

Chapter Twenty-Four—The First Mass-Produced Front-Wheel Drive is in the Cord in 1929

Three of America's most elegant autos at one time came from the same place—Auburn, Indiana. The three cars were the *Auburn,* the *Duesenberg*, and the *Cord*. In 1900 Charles Eckhart and his two sons Frank and Morris, like many others, started making carriages and graduated to motor vehicles.

It started when Frank, who served as the salesperson for the company, went to Lansing, Michigan and purchased a *Curved-Dash Oldsmobile*. Rebuilding it to their liking, their first cars were one-cylinder models; in 1905 they started building two-cylinder cars.

The business was sold to a consortium of Indiana businessmen in 1912 and the new owners came out with a six-cylinder auto. In 1924 Erret L. Cord became general manager and in

1925 introduced eight-cylinder autos to the firm. In 1929, the auto named the *Cord L-29* in his honor became the first front-wheel drive car offered to the public—an option that just about every car company would introduce and make standard equipment. Models called the *810* and the *812* had the first "hidden headlights."

The 1929 Cord L-29 Cabriolet

In 1937 the *Cord* ceased production and was sold to **Aviation Enterprises**. E.L. Cord moved to Nevada, where he made millions in real estate!

1937 Cord Phaeton

Part Four-

The Classic Era-

1931-1948

Chapter Twenty-Five—The First All-Mechanical Automatic Transmission is Developed by REO in 1932

There are a lot of contenders for the first automatic transmission. The most cited is **REO**, the motor company Ransom E. Olds started after leaving the **Olds Motor Works**. **REO** was a very successful company over the years. It didn't take them long to overtake **Olds Motor Works**, the first company Ransom Olds started.

In 1932 **REO** developed the first mechanical automatic transmission in a mass-produced vehicle. They advanced into trucks and manufactured the famous *REO Speedwagon*. Many people don't realize they were also one of the longest-lived companies, lasting long into the 1970s. `

The last REO factory in 1977

Chapter Twenty-Six—Packard Introduces Air-Conditioning in 1939, Power Windows in 1941, and Power Locks in 1954

When Detroiter Henry B. Joy saw two Packard owners take off like a bolt of lightning at the New York Auto Show, he was hooked. Buying one of his own, he headed down to Warren, Ohio, where James Packard and his brothers Hatcher and Weiss were manufacturing their much-admired auto. Joy talked the brothers into moving their whole operation to Detroit, and a dynasty was born. **Packard Motor Company** became known as the producer of some of the most beautiful and innovative luxury cars in the business.

In 1939, air-conditioning units were provided by **Packard** as an option for $274. The units, designed by the Bishop and Babcock Co. of Cleveland, Ohio, could also be set to provide heat. The units had a lot of flaws and were discontinued by **Packard** in 1941.

Power windows (also known as "electric windows") made their first appearance in the 1940 *Packard 180 Series*, in a "hydro-electric" system. **Packard** customers had 16 years where they could electrically roll their windows up and down but still had to manually lock their cars. Finally, in 1956, this was rectified as **Packard** became the first car company to offer power locks and power windows in their vehicles. **Packard** would have been the first with power locks, but the **Scripps-Booth Auto Company** had them first in one of their cars in 1914.

A Los Angeles, California *Packard* dealer had the first neon signs in the nation—and they worked, stopping traffic!

Alas, the good times were ending by the 1950s. In a battle for the hearts and minds of the American family and what they want in a family car, **Chevrolet** and **Ford Motor Companies** began lowering their prices. This was hurting the car companies that had made it to the 1950s, nicknamed "the Little Four," **Hudson, Nash, Packard,** and **Studebaker.**

Studebaker felt they could still make it on their own, but **Hudson** and **Nash** formed an alliance that became the **American Motor Company**. **Packard** had been invited to join, but when the new consortium wanted to put **Nash**'s George Mason in charge, and not **Packard**'s James Nance, Nance quashed the deal. It was just a year after that **Packard** was so cash-poor that it was forced to make an alliance with **Studebaker**, who moved the company to South Bend, Indiana from Detroit, and then retired the marque in 1959.

Chapter Twenty-Seven—Nash Had the First Seatbelt in a Mass-Produced Car in 1948

Autos had been around a long time in 1948, and the number of fatalities from drivers and passengers going through the windows or getting thrown from the car due to an accident (or just a fast stop) was increasing each year. Seat belts had long been used by racecar drivers but weren't offered to the general public until 1948, when **Nash Motors** became the first auto company to finally include seat belts as an option. There wasn't another company that provided seat belts as an option until seven years later, when **Ford Motor Company** offered them in 1955.

Nash Motors began when former GM President Charles W. Nash purchased the **Thomas Jeffrey Company** of Kenosha, Wisconsin in 1916. The most famous car of the **Thomas Jeffrey Company** was the *Rambler*, and **Nash** brought out their version in 1917, the *Nash Model 671*. **Nash Motors** had many successful years under their founder Charles Nash. Nash orchestrated a merger between the **Kelvinator** appliance company and **Nash Motors** in 1937, right before naming his successor, George W. Mason, and retiring to his Beverly Hills, California home.

The newly-christened Nash-Kelvinator Corporation changed to war production during the 1940s. After World War II was over, Nash made the transition back to automobiles. It was a difficult time for the remaining auto companies. **Graham-Paige** was bought by **Kaiser-Frazier**, and **Nash** and **Hudson** joined forces to form the **American Motor Corporation (AMC).**

Chapter Twenty-Eight—Crosley Had the First Mass-Produced Disc Brake in 1949

In 1949, the name Crosley was usually associated with radios and electronic equipment. However, not only did the **Crosley Corporation** come out with a car but also had the first disc brakes in an auto!

Powel Crosley, Jr. started the corporation named for him, along with his brother Lewis. They started building inexpensive radios and radio kits and did very well. They were heavily involved in early broadcasting, and owned baseball's Cincinnati Reds.

Since the dream of Powel was always to go into auto manufacturing, in 1939 the **Crosley Motor Company** was born. The first autos were smaller and were priced competitively. The company had a good start, interrupted by World War II, when they went into war production. After World War II, they returned to auto production and manufactured the first post-war sports car, the *Hotshot*.

The 1949 Crosley Hotshot

In 1949 the Crosley manufactured an auto with two-caliper, disc brakes, the first in the auto world. But the revolutionary brakes did not stop the Crosley's slide into obscurity. As the 1950s dawned, Crosley sales were going down, until finally, in 1959, the company went out-of-business.

1949 Crosley sedan

Part Five

The Modern Era

1949-Present

Chapter Twenty-Nine—The Plymouth Valiant Had the First Alternator in 1960

The first alternator in an auto was in the **Chrysler Corporation**'s 1960 *Plymouth Valiant*. The alternator charges the battery, and powers the electrical system for the car's accessories. Before alternators, cars used direct current (DC) generators, outfitted with commutators which would switch the current from the rotor to the external circuit. Today's automotive charging systems include a battery, voltage regulator, and the alternator. The name alternator comes from "alternating current."

As autos needed to support more options, including power windows, locks, large headlights, electric windshield wipers, and defoggers (heated rear windows), it needed a better way to generate power for all the various power accessories. **Chrysler** beat the other two members of the "Big Three," **Ford** and **General Motors**, by a few years with its alternator.

The first car with an alternator, the 1960 Plymouth Valiant.

The *Plymouth Valiant* was Chrysler's entry into the "compact car" field of the early 1960s and was very successful, selling a lot of cars. It was said that the *Valiant* kept **Chrysler** afloat during some of the lean years. *Road and Track Magazine* called the *Plymouth Valiant* "one of the best all-around domestic cars."

The 1969 Plymouth Valiant Signet 2-door Sedan, one of many Valiant styles

By 1975 the Valiant brand was morphing into several offshoots, including the *Plymouth Barracuda, Plymouth Volare*, and the *Plymouth Duster*. By 1977 it was gone.

The 1975 Plymouth Valiant Brougham, one of the last of the brand.

Chapter Thirty—The *Glasspar* From California Had the First All-Fiberglass Body in 1950

In 1950, the *Glasspar* of Santa Ana, California manufactured the first fiberglass car, which was a sports car roadster model. The **Glasspar Company** manufactured 200 cars from 1950 until they quit in 1955. The fiberglass bodies were "lightweight, rustproof, dent-resistant, and easy to repair."

They were also inexpensive to produce--$950 per body. The European-styled cars were often outfitted with a *Lincoln* engine. When **General Motors** wanted to manufacture the fiberglass-body *Corvette Sting Ray*, they consulted with fiberglass expert and *Glasspar* designer William Tritt.

A Glasspar from California

Part Six-Orphan Auto Companies by Region

Chapter Thirty—Orphans of The South

One of Alabama's few attempts to get an automobile company started was in 1922, the **McCormack Brothers Motor Company** was organized in Birmingham, Alabama and assembled a steam vehicle, but it never went beyond the prototype stage.

A Georgia auto company was the **Primo Motor Company,** organized in 1910 by E. Van Winkle of the **Van Winkle Gin and Machine Company**. The *Primo* automobile was a 4-cylinder, 25 horsepower vehicle, available in several styles, but unlike the more successful Hanson and White Star companies of Georgia, quickly faded away.

Another similar company was the **Southern Automobile Manufacturing Company**, founded in Jacksonville in 1906. A few prototypes of the "high-wheelers" were said to have been built.

The 1952 Cunningham-Briggs

Starting in 1951, the **B.S. Cunningham Co**. built a car commissioned by race car driver Briggs Cunningham. The car was built in West Palm Beach, FL and was available as a coupe or a roadster, with the choice of a 200 hp or a 310 hp engine.

Although considered a sophisticated design, the cars were costlier to produce than what they could be sold for. They made their exit from the market in 1955.

In 1911, The **Orr Modern MotorCar Co.** was incorporated at $2 million dollars in Yazoo City, Mississippi. Their prototype was the *1911 Orson Touring Car* with the special feature "the worm drive." Evidently it never caught on since they weren't heard from in 1912.

In 1920, Dr. A. Richard Carter of Hammond, Louisiana moved his **Richard Carter Motor Co**. to Gulfport, Mississippi. They manufactured the 1920 *Carter Steam Car*, which built up steam by spraying kerosene on a coil boiler.

The 2-cylinder vehicle would get a reported 20 miles to a gallon of kerosene. Approximately 25 Carter Steam Cars were manufactured and sold for $2,350. Reputedly about to manufacture steam tractors, the company had faded away by 1921.

Other reported Mississippi auto companies and cars include the *Aberdeen* from Aberdeen, MS., and the *Waltham Steam* from Waltham, Mississippi. Scant information is available on these two cars and companies.

The **Victor Motor Car Company** was one of two making the Victor cyclecar/runabout in 1913. Under the auspices of C.V. Stahl, they built a factory in Greenville, South Carolina, and produced a touring car until 1917, when poor sales forced them to shut down.

Another Greenville, S.C. company was **Cyclone Motors Corporation**. Although they mainly made trucks, in 1921 they built a fleet of black and white taxicabs for New York City. The cabs were called *W.B.C.s,* standing for "Well Built Cab." Although they incorporated with one million dollars in capital, by 1922 they were over $300,000 in debt and liquidated the company.

Chapter Thirty-One: Orphans of The Far West

From 1910 until 1913, the **A Automobile Company** tried to launch their company and their car, the *Blue and Gold,* named for the California state colors. The Sacramento, California company was given land for their factory, with the stipulation that the building of a car factory would start soon. In 1911, they established offices in downtown Sacramento and promised that the car factory would be built soon.

The company built a couple of prototypes which they exhibited in 1913 in a Sacramento dealership showroom. There was a 4- and a 6-cylinder, both with left-hand drive, electric lights, and self-starters as standard equipment. Advertisements ran in the *San Francisco Examiner* advertising the 4-cylinder car for $1,150 and the 6-cylinder model for $2,100.

Few cars were built. The company tried to get concessions from Richmond, California to move there, but a lack of interest caused the company to liquidate in 1913.

The *Macomber* was built by the **Macomber Motors Co.** of Los Angeles. It was owned by Walter G. Macomber who was a designer of rotary engines used in the *Eagle* and other marques in 1917. He became known as the inventor of the "vee-radiator." Only a few *Macombers* were built.

In 1914, George, Frank, and Charles Parker incorporated the **Parker Motor Car Company** (later briefly known as **Ajax Motor Company**) in Seattle, Washington, to manufacture their *Ajax* and *Ajax Six* auto. They had the gears, crankshafts, and axles forged by **Krupp** in Germany and had their own foundry to make bronze, aluminum, and iron components.

The cars were available with a 4- or 6-cylinder engine and the customer could get a sleeve valve instead of a poppet valve on his 6-cylinder "for a reasonable cost," the substitution being made while the car was on the assembly line. But the company never made it all the way through 1915, their second year.

The **Eureka Motor Company** began in Seattle in 1906 and produced a 4-cylinder, 24-horsepower, two-passenger runabout. They purchased the **Seattle Manufacturing & Supply Company** at 1409-13 Broadway in Seattle in 1907 and started to build a two-story factory. But pricing their 2-passenger runabout at $1,900 was just setting themselves up for failure, and they did, later that same year.

Philo E. Remington was the grandson of Eliphalet Remington, the man who founded the Remington firearms and typewriter companies. Philo wished to start an auto company and formed the **Remington Standard Motor Company**. He brought the company to Charleston, West Virginia where they produced autos from 1910 to 1913.

Classic car aficionados are well-aware of the *Tucker,* a 1948 auto of Preston Tucker that had a well-regarded movie made about it. But few are aware of a different *Tucker* made from 1900 to 1903 by William Tucker in California. This *Tucker* had a 2-cylinder, air-cooled engine with artillery wheels. It was estimated that about 16 were made.

The *Hartman*, named for inventor George Hartman, was produced by the **Model Gas Engine Works** in Red Bluff, California and had a 110-inch wheelbase, powered by a 4-cylinder gasoline engine. From 1914 until 1918, twenty cars were manufactured, and then George Hartman went into the Army. When he returned from his tour-of-duty, he didn't return to making cars.

The *Davis* was a three-wheeled auto designed by Gary Davis in Van Nuys, California from 1947 to 1949. There were about 17 to 100 of the prototype auto manufactured, although none were for sale. The cars had a 4-cylinder Hercules engine, all-aluminum tubular body, and Kinmount disc brakes. Later designs had Bendix brakes and a 4-cylinder Continental engine.

(Continental and Lycoming engines were the two biggest supplier of auto engines to independent auto companies.)

Although the **Davis Motor Company** advertised that the car could go 100 miles per hour, and would get 35 miles per gallon, the real numbers were closer to 75 mph and 28 mpg. The brand never found enough investors to mass-produce the auto.

Chapter Thirty-Two—The Rocky Mountain State

Coleman Motors of Littleton, Colorado, built a low-priced, under $1,000 car with the engine mounted under the front axle. The 1933 auto had straight body sides and didn't include mudguards. The car failed to enchant the public, but probably not just because of the lack of mudguards.

In 1903, the **Cushman Motor Company** produced the *Cushman Runabout* in Lincoln, Nebraska. Although they attempted to raise $300,000 in a stock offering, they only got $50,000. It is surmised that six sales were to family members, since the *Nebraska Motor Vehicle Registry* for 1903 shows that six of the Cushman owners had the last name Cushman. They probably produced no more than 10 of the vehicle, although four more were registered in Nebraska (other than the ones that were family-owned).

Chapter Thirty-Three—Orphans of the New England States

The **H.H. Buffam Company** produced the *Buffam* from 1901 until 1906, in Abington, Massachusetts. Their first cars had 4-cylinder engines, but in 1904, their *Model G Greyhound* came out with the first 8-cylinder engine in the U.S. In 1905, their cars switched to a V-8 engine.

Besides the cars, the company also manufactured motor boats. Buffam's design of the aluminum body brought the car in at only 120 pounds!

Following the demise of the *Buffam,* founder H.H. Buffam re-entered the automotive field in 1914 by buying and producing the *Laconia* cyclecar.

In Holyoke, Massachusetts, the **Holyoke Motor Company** built a 2-cylinder, 7-horsepower vehicle in 1899, designed by Charles Greuter. He started **Greuter Motor Works** in 1899 and as more people wanted in, he reorganized as the **Holyoke Automobile Company** in 1900.

After manufacturing vehicles until 1903, the company was purchased by the Matheson brothers and became the **Matheson Automobile Company,** manufacturing the same car with a different nameplate. Greuter stayed with the Mathesons until 1908, when he left to produce his own vehicle, the *C.R.G.,* named for his initials. From there he went to the **Excelsior Motor Manufacturing Company** of Chicago, IL, where his work was seen by the heads of **Stutz Motor Company**. They recruited Greuter and he was with **Stutz** until the end.

In Worchester, Massachusetts, the people who worked at the Crompton Loom decided to go into the automobile business in 1901. They constructed two distinct types of car, a steam car and an electric. Calling themselves the **Crompton Motor Carriage Company**, the founder, Charles H. Crompton, was President.

Deciding that steam would be the way to go, they released their first car to the public in 1903. It had a unique system to use the steam to power the vehicle. They sold a few at the Boston Automobile Show in 1903 but not many. When the factory in Worchester caught fire in May 1905, the company called it quits.

———————————

"Made and Tested in the Berkshire Hills" was the slogan of the Pittsfield, Massachusetts **Berkshire Motor Company.** Their 1905 *Berkshire Model A* had a 4-cylinder, 18hp engine, and the auto had 86-inch wheelbases up to 122 inches wb, depending on the model chosen. With only a few cars built, they shut down in 1907 until, as they announced, they could acquire more capital.

In 1909, they resurfaced and built approximately 30 vehicles for 1910. Relocation to Cambridge in 1912 only added about three more cars to the total of 150 total autos built during the life of the auto, which ended in 1912.

The **Sultan Motor Company** produced an auto named the *Sultan* in Springfield, Massachusetts, from 1906 until 1912. At first the company produced commercial vehicles like taxicabs, but in 1908 decided to have a car offered to the public. When the taxis were the main line, they were licensed out to **Elektron Manufacturing Company,** who thought they would be producing the public offering. But instead, the contract was given to the **Otis Elevator Company! The Otis Elevator Company** built the *Sultan,* a 4-cylinder, 3-speed auto. By 1912, with the proliferation of skyscrapers, the Otis elevators became more popular and took up all of the company's time, so the *Sultan* motorcar ceased production.

From Greenfield, Massachusetts, Max Hertel was an entrant in the 1895 *Chicago Times Herald* Contest with the smallest gasoline car entered. It was a 2-cylinder, 2 hp gasoline-engine vehicle that Hertel, an engineer with the **American Biscuit Company**, designed himself. By 1899, he was manufacturing and marketing a 500-pound two-seater runabout for $750. The public didn't care for the bicycle-like frame of the vehicle and he was out of business by 1900. He printed way more catalogs than he ever produced cars.

1910 Touring Model Sultan and the 1910 Landaulet Sultan

From 1926 until 1942, Harvard and M.I.T. graduate Thomas Derr of West Newton, Massachusetts, operated the **American Steam Car Company**. He created custom steam cars from various other vehicle parts, such as Hudson bodies and his own "V-4" design Stanley steam engines. His cars were known as *American Steam Cars* and he produced at least 50 to 100 of them until 1942. From 1942 until his demise in 1948, he worked on developing an artificial fog for the U.S. Army.

In 1904, George Hill founded the **Hill Motor Car Company** at 108 Merrimack Street in Haverhill, Massachusetts. He produced a vehicle called the *Hill,* which was an $1,850 2-cylinder touring car with an air-cooled engine. However, by 1908 only ten cars had been produced and Hill suspended further production. The chief problem had been "aluminum pins that sometimes fell out," said his widow in a 1950s interview.

In 1919 until 1920 in East Warren, Rhode Island, the **Greyhound Motors Corporation** built a roadster and a touring car with 30 hp, 4-cylider car with a 106-inch wheelbase called the *Greyhound.* Not many were made.

In Hartford, Connecticut, in 1909 the **McCue Company** decided to move beyond manufacturing carriage fittings and produce their own auto. The car they built was called the *McCue* and to build it the company moved from its plant on Capitol Avenue to larger quarters on Pliny Street.

The car was a medium-powered 4-cylinder car, with an L-head engine, available in 30hp or 40hp, in a medium price range. The cars had 3-speed transmissions with shaft drives.

One of the models was the *Gentlemen's Roadster.* The car didn't sell well, and by 1911 the company had merged with **Superior Axle & Forge Co.** of Buffalo, New York, to specialize in making car axles.

The *Skene* was a 2-cylinder, 5 hp steam car built in Lewiston, Maine. In 1900, the company claimed they had already built 125 cars and was looking for investors. In 1901, nothing more was heard from the fledgling company.

In March 1914, the **Maryland Electric Vehicle & Manufacturing Company** was organized in Baltimore for the manufacture of 10,000 lb. capacity electric vehicles. Although few of the larger vehicles were built, the company built a few small, 1.000-pound vehicles. These sold for $1,250. The company went out of business the same year they started.

A vehicle called the *Maryland* in 1922 out of Frederick, Maryland never made it out of the planning stages. Before the firm could finish its prototype in July 1932, the company had gone into receivership.

The **Scott Iron Works** of Baltimore built several custom cars from 1901 until 1904. One of the most notable was the one built for Harlan W. Whipple, the president of the American Automobile Association. The auto had a 4-cylinder, 80 hp engine, and was in the style of a massive touring car, seating seven, with a 126-inch wheelbase.

The **Malden Automobile Co**. organized in Malden, Massachusetts in 1897 and in 1898 put out a light, two-seater steam car named the *Malden*. Two extra seats could be fitted to the car. It had a vertical 2-cylinder engine and wire wheels.

In Brookton, Massachusetts, brothers Emil and Gerard Pickard started repairing bicycles in 1896. In 1900 they added auto repair to their list of services. They started experimenting with building their own auto in 1903, and by 1908 the brothers thought they had a product that was marketable.

They released their car, the *Pickard,* in 1909, with a 4-cylinder, air-cooled engine, shaft-drive, and sliding-gear transmission. The motto was, "A Lot of Car for the Money," and most of their vehicles sold for less than $1,000, although some were sold for as much as $1,500.

The last year of the marque was 1912, when the Pickard brothers released a statement stating that they were "handicapped by lack of capital."

The **Lenox Motor Car Company** started in Jamaica Plain, Massachusetts, in 1911, succeeding the **Martell Motor Company**, which had tried manufacturing a vehicle but failed. The first *Lenox* was a 4-cylinder, 27-horsepower, vehicle which was available in runabout, touring car, and speedster styles, and even a limo model.

In 1912 an additional plant in Hyde Park, Massachusetts was added. A 6-cylinder, 40- and 60-horsepower line of cars was added in 1913.

When the company wanted to manufacture commercial vehicles, another plant was secured in Lawrence in 1915. The new head was Daniel Emerson, of the **Emerson Shoe Co**. and in 1916 the company introduced a six-ton truck and a few tractors.

The company spent too much retooling the factories to manufacture commercial vehicles and in 1917, the line shut down, not to be re-started.

The steam boiler factory of Edward S. Clark in Boston, Massachusetts led to Clark experimenting and building a steam car in 1901, powered by a 4-cylinder, 20hp engine. It featured undoubtedly the first "tilting steering wheel" for easy entrance and exit to the vehicle. Called the *Clark Steam Car,* the price was originally $5,000, but in 1909, this was halved to $2,500.

The car never sold in mass quantities and was produced in a variety of models and styles. By 1909, Edward Clark was no longer building steam autos but was dabbling in gasoline-powered trucks. In later years, he continued manufacturing and furnishing auto parts, as well as performing regular auto repairing, remodeling, and painting autos of all types.

In 1893, the *Simonds Steamer* of Clarence Simonds was built in Lynn, Massachusetts.

With a top speed of 10 m.p.h. and a short, 5-minute warm up time until the steam could cause combustion, the car was one of the first in the U.S. to be a viable steam vehicle. Simonds only used the car to make deliveries on the streets of Lynn, but he did talk to the Stanley twin brothers, Freelan and Francis of Newton, Massachusetts, about steam car technology. Clarence never built any other steam cars but advanced steam car technology when he advised the Stanley twins.

_____

The *Steamobile* was a 9hp, 2-cylinder, single-chain drive vehicle built in Keene, New Hampshire from 1900 until 1902. A product of the **Keene Automobile Company,** it used a tiller for steering and was available in two-seater and four-seater models.

The **Roader Car Company** of Brockton, Massachusetts, was a short-lived car company that manufactured a 4-cylinder runabout called the *Roader* (slang for a fast horse) in 1911 and 1912. The auto had a sliding gear transmission, and its styling included a round gasoline tank in the rear and a "dropped frame."

Two different horsepower options were included—either 20 horsepower or 30 horsepower. In 1911, only the 20 hp option was offered and the company went out of business at the end of the year.

———————————

The *Eclipse Steam* vehicle of Easton, Massachusetts, was built by the **Eclipse Motor Company** from 1901 to 1903. It was a coach-like vehicle with a 62-inch wheelbase, wire wheels, and tiller steering. The sold seven autos at their first auto show in Boston, including one to the U.S. Post Office. However, the Post Office didn't decide to order a fleet of vehicles, and the undercapitalized **Eclipse Motor Company** was out of business by 1903.

The designer of the *Eclipse Steam* was Everett S. Cameron, who went on to build the air-cooled, self-named *Cameron*.

The **Med-Bow Automobile Company** was named by Harry Medcraft and George C. Bowersox. The *Springfield* was a 4-cylinder, 35 hp touring car with a 107-inch wheelbase, manufactured in Springfield, Massachusetts from 1907 until 1909. The factory moved to Springfield, Illinois in 1909 to produce *Springfields* in Illinois until 1910. Total production in

Massachusetts was probably about 46 cars, while the reported production in Illinois was anywhere from 11 to 200.

The **Heymann Motor Vehicle and Manufacturing Company** of Melrose, Massachusetts, was capitalized at $250,000 in the summer of 1898. The original prototype was a 3-cylinder, 4-cycle, 6 hp, water-cooled engine-powered vehicle with tiller steering. Although 50 cars were reported to be "in the building stage" in 1899, by 1900 it was reported that the company was being reorganized with a capital stock of $350,000.

Little was heard until 1904, when a new prototype appeared—a 5-cylinder, 40 hp, rotary-engine powered touring car. It contained the "Gearless Variable Speed Controller," which was a variation of the friction transmission.

Once again, no reports of actual production on the vehicle were reported. The vehicle made a swan-song appearance at the 1907 Boston Automobile Show and thereafter disappeared entirely.

The **Lorraine Motors Corporation's** *Lorraine* had a 4-cylinder Herschel-Spillman engine and was available in both open and closed-models. But production never went beyond a few hundred cars, and they were out-of-business by 1922.

The Lorraine Auto

Chapter Thirty-Four—Orphans of The Midwest States

The **Lewis Motor Company**, for William Lewis, was changed to **L.P.C. Motor Company**,

for the initials of the other founders, Rene Petard and James Crarn. All three were alumni of the

Mitchell Motor Company.

From 1914 until 1916, **L.P.C. Motor Car Co.** produced cars in Racine, Wisconsin, using

one of America's first long-stroke engines. The car was called "the Monarch of the Sixes," in

reference to its 6-cylinder engine. The Vulcan electric gearshift was introduced on the **L.P.C.**

vehicles.

At first the *Lewis* was only offered to the public as a six-passenger touring car. In 1915, a roadster version was added.

World War I spelled the demise of the *Lewis* automobile as many key employees were drafted. The company paid all 280 workers and relocated them to other jobs in Racine, Wisconsin. After the auction of the company's assets, all its creditors were paid off.

Brothers Charles and Frederick Piggins built their first steam car in 1883, their first electric in 1897, and their first gasoline vehicle in 1902. Their machine shop at 1113 Sixth Street in Racine, Wisconsin was making 2-, 4-, and 6-cylinder engines that were, according to the advertising, "smokeless and noiseless."

The first *Piggins* automobiles came out in 1908 by the **Piggins Brothers,** and had T-6 engines, available with 36 horsepower at $3,500, and with 50 hp at $4,700. The motto was: Piggins: The Name that Stands for Perfection in High-Class Auto Building. Ads also emphasized the "luxurious beauty" and "splendid efficiency" while emphasizing how quiet the motor was.

In 1910 the company consolidated with the **Racine Motor Company** and quit passenger car production to become the **Piggins Motor Truck Company.** They built a truck called *"The Practical Piggins"* that was popular and was manufactured until 1916. The company then made the truck called the *Reliance* which continued into the mid-1920s.

In 1903, a Lansing car maker, the **Olds Motor Company** manufactured the top-selling auto in the U.S.: the "merry" Oldsmobile called the *Curved Dash Olds.* **Olds Motor Company** held the top spot until 1906, when **Ford Motor Company** of Detroit, Michigan took over the number one spot and held it until 1927, when Flint's **General Motor**'s **Chevrolet Motor Company** knocked **Ford Motor Company** down to number two.

Oldsmobile was not the only car company in Michigan's capital city. Lansing was also home to the *Nu-Klea Starlite*. The *Nu-Klea* was an electric car built by the **Nu-Klea Automobile Corporation** in 1959 and 1960. The car weighed 2,100 pounds and had a plastic body with TWO motors driving the rear wheels. The name was apparently a play on "nuclear."

 *The Nu-Klea Starlite from Lansing, Michigan*

Emil A. Nelson founded the **E.A. Nelson Motor Car Co**. to produce his 4-cylinder *Nelson,* a touring model built with an overhead camshaft in Detroit. He produced a few roadsters and closed models also, about 350 cars in all, from 1917 until 1921. Nelson had worked for **Oldsmobile, Hupp,** and **Packard Motor Companies** prior to starting his own company.

About 400 *Nelson Roadster*s were built from 1917 to 1922. With a 4-cylinder overhead engine, it was a lightweight car that could go fast yet was good on fuel consumption—25 miles to the gallon, which was good for the Brass Era autos. They sold in the $1,500 range.

1917 Nelson Roadster

In the Upper Peninsula of Michigan, the **Menominee Electric Manufacturing Co.** built the *Menominee* electric automobile in 1915. Producing mostly commercial vehicles, the company also manufactured a cabriolet model. Its top speed was 20 mph, and it could go 50 to 60 miles on a single charge. With a price tag of $1,250, a recharging kit for the battery was included. The auto marque lasted barely a year.

In Grand Rapids, Michigan, the **Berwick Auto Car Company** manufactured the *Berwick* in 1904, an electric car. It was a two-seater runabout with three-speeds and steered by a tiller. The reasons for the company's failure was probably the high price of the vehicle--$750, and the low top speed of the vehicle—20 mph.

In 1906, the **C.V.I. Motor Car Company** was founded in Jackson, Michigan with $100,000 capital stock. In December 1907, the 6-cylinder, 40-horsepower car was introduced in the Chicago Auto Show. In January 1908 a factory was purchased and outfitted to make the *C.V.I.* auto. Designed by engineer Charles Cutting, *C.V.I.* stood for "Cutting VI" (6 in Roman Numerals).

The 1908 *C.V.I.* was available in a touring and runabout model. Although the car was very well-received by car aficionados, the original backers backed out, and Cutting had to look for different investors. That was the last heard.

_____

The factory that was left empty when the *Aerocar* went out of business was rented by a brand-new company called the **Hudson Motor Company** who used it and then moved to larger quarters on Jefferson Avenue Other car companies used it afterwards. The factory was used by, In order:

1. *Aerocar 1906-08*

2. *Hudson 1909-10 (moved to Jefferson and Conner)*

3. *Everitt 1910-12 (auto, not part of the coach business)*

4. Columbia 1916-23 (bought Liberty in 1923, moved to Conner & Charlevoix near the Hudson plant).

The factory at Mack and Beaufait in Detroit today.

———————

The **Lewis Spring & Axle Company** of Jackson and Chelsea, Michigan, operated from 1915, until 1921, manufacturing the open-model *Vincent-Hollier*. The company built their own V-8 engine for the auto. In 1917, they started offering a second vehicle with a 6-cylinder Falls engine until their demise in 1921.

1915 *Vincent-Hollier Eight*

The **Motor Buggy Manufacturing Co.** of Minneapolis, Minnesota, released its *Acme Roadster* in 1908. A 1,600 pound, 2-cylinder, 22-horsepower 4-passenger vehicle capable of doing 25 miles per hour, the back seat could be removed, allowing the car to haul up to 500 pounds.

When the car's sales were deficient, in 1909 the car's name was changed to the *M.B.* and in 1911, a 4-cylinder vehicle was added. After 1911, the company concentrated on commercial vehicles.

———————

An early steam car was the one built by Lambert Kemp in 1893. It used wind to power the boiler and was described as just a board with an engine.

The *Imperial* was built in Columbus, Ohio, by carriage maker **Rodgers & Company** in 1903 and 1904. They had coupe, runabout, and light delivery models with 2-cylinder, 8-horsepower engines and 78-inch wheelbases. In 1904, a surrey model was added.

Production was small and ended completely by 1905.

The *Jewell* was built by the **Forest City Motor Car Company** of Massillon, Ohio in 1906. The car was a 1-cylinder highwheeler with rope drive. In 1908, a four-cylinder model was added to the line and an "I" off the car name, so that it was the *Jewel.* In 1909 the company merged with the **Croxton-Keeton Motor Company**.

The **W.H. Gabriel Carriage & Wagon Company,** established in 1851, wanted to break into the automotive industry. In March 1910, they exhibited their 4-cylinder, 30-horsepower touring car with a 120-inch wheelbase at the Cleveland Automobile Show.

As the **Gabriel Motor Company,** they had offices located at 438 Broadway in Cleveland and a factory at 1674 E. Third Street. They manufactured their auto until 1912. In 1913, they manufactured a one-ton *Gabriel* truck which proved successful enough for them to continue with the trucks until the end of World War I.

In 1891, John W. Lambert owned a lumber yard, grain elevator, hardware store, the local opera house, jail, and the town hall of Ohio City, Ohio, right across the border from

Anderson, Indiana. The same year he successfully tested a 3-cylinder, gasoline powered surrey of his own design that some consider the first gasoline-powered vehicle in the U.S.

He organized the **Buckeye Manufacturing Company** in Anderson, Indiana, to build gasoline engines. In 1895, he released a three-wheel vehicle he called the *Buckeye.* In 1898, he developed a four-wheeler with a friction transmission.

A Lambert from Union City, Indiana.

He achieved success in 1902 with a car built in Union City, Indiana called the *Union.* The 2-cylinder car was #8 in auto sales in 1902. Most of the components of the auto were assembled in Lambert's Anderson, Indiana factory and by 1905 the production of the *Union* was moved there also.

In 1906 the production of the *Union* was discontinued, and the car called the *Lambert* made its debut. The smaller, two-cylinder models were chiefly *Union*s with the name badge changed. The smaller *Lambert* was joined by a 4-cylinder model.

About 200 *Lamberts* were manufactured per year until 1917. The company continued to make trucks, which they had started doing in 1900, but totally ceased vehicle production the next year.

———————

The **Chicago Coach & Carriage Company** of Chicago, Illinois built a two-seater, high-wheeler vehicle in 1907 and 1908. The vehicle, called the *Duer*, had the 2-cylinder, 15 hp gasoline engine located under the car's seat in the 1907 models. For 1908, the engine was moved under the hood.

1907 Duer High Wheeler

Like many companies, they started out making carriages and switched to cars. Probably the profession to have the most carmakers were the companies that previously made buggies and carriages. Once cars started to become popular, carriage makers could see that cars would eventually put carriages out of business. The fact that carriage-makers had large work areas was also a factor for many companies to switch to car production.

The **Available Truck Company** of Chicago made the *Available Truck* from 1910 to 1957. They would manufacture 150 trucks per year, for a total of 2500 over the years. The truck shared space with the **John Rath Cooperage Company** at 2541 Ellston Street in Chicago. John Rath was also the owner of the **Available Truck Company.**

Most of the trucks were sold to Chicago customers. In 1957 the company was sold to **Crane Carriers Corporation** of Tulsa Oklahoma.

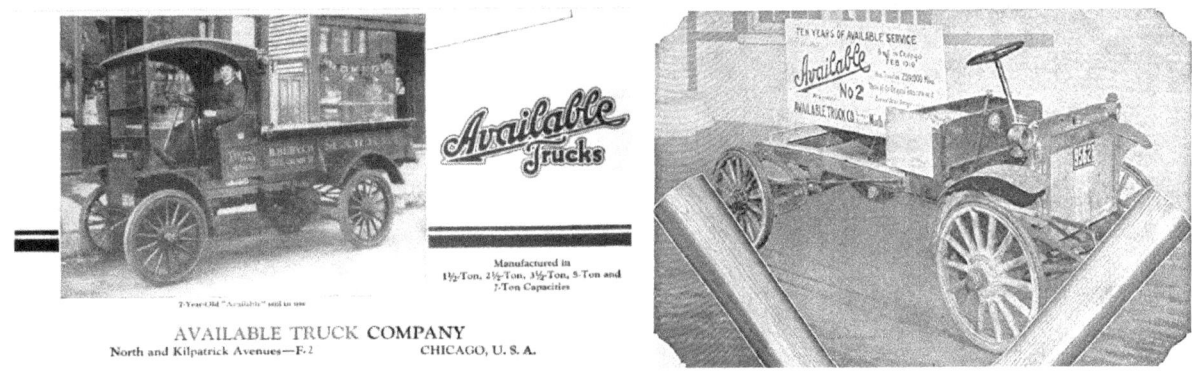

Available Truck Company advertisement, display of the 2nd Available "truck."

Commercial Car Magazine ad of 1944 for Available Trucks

The **American Motor Car Company** of Indianapolis, IN started building the *American Scout* in 1906. The *Scout* was a 4-cylinder, 20hp vehicle which had the axles placed above the frame. This became the "underslung" design and helped cars navigate rough roads.

The *Scout* had a compression starter that pumped air into the cylinders, eliminating the need to crank the car more than once. This preceded the Kettering starter, first used on the *Cadillac*.

This **American Motor Car Co.** was not related to the **American Motor Company** (AMC) that later formed in Detroit.

The 1913 American Scout of the American Motors Co. of

Indianapolis.

In the spring of 1905, J. Jeffrey, J.M. Spiker, and E.D. Pinney decided to transform their **Aurora Carriage Top Company** business into **Aurora Automobile Company** and assemble automobiles. W.H. Howe of the **Chicago Motor Company** designed a 4-cylinder, 30 horsepower touring car with a 100-inch wheelbase, inventing a new friction transmission for it. The marque lasted until 1906, one year.

The *Pilot*'s motto was "The Car Ahead." Starting in 1909 in Richmond, Indiana, the **Pilot Motor Company** outlasted most Indiana companies by continuing to manufacture autos until 1924. It started, as many car companies did, as a carriage company. It was named the **Seidel Buggy Company** for owner George Seidel. The cars were named *"Pilots"* because George had always wanted to be a river boat pilot.

The cars were at first built in the buggy factory, until they moved across town to a new, 500-car-a year capacity auto plant in 1910. The *Pilot* was at first a 4-cylinder roadster. In 1913 a 6-cylinder was introduced and in 1917 all the *Pilots* were 6-cylinders. The only year they offered a V-8 engine was in 1916. One of the most popular models was the 1922 *Pilot Sportster* with barrel-type headlights and no running boards.

George Seidel was proud of the fact that he was one of the first auto plants to hire women (even though he only had them sew finish upholstery and curtains).

The post-World War I depression was the main reason for the *Pilot*'s decline into receivership in 1918. George Seidel remained president of the company during the company's entire 15-year run (longer, if you count the buggy years). He blamed the "cut-throat tactics" of the "Eastern money interest" for the company's demise. Over 4,000 *Pilots* were built during the life of the marque.

———————

The *DeTamble* auto started life in Indianapolis in the factory of the **Speed Changing Pulley Company.** Edward De Tamble was the head of the company that produced the two-cylinder runabout in 1908. In August 1909 the company changed their name to the **DeTamble**

Auto Company and moved production to a factory in Anderson, Indiana. In 1910, the company started manufacturing a 4-cylinder auto.

In 1911, the company had a financial crisis and DeTamble was out as president and a new board composed of creditors and debtors of the company was seated. By this time, over 2,000 *DeTamble*s had been sold. In 1912, new president Charles Walters was arrested for embezzling $16,654 from the company but acquitted. By 1913, the company stopped producing cars and in 1915, the machinery was sold.

"The lowest priced successful car on the road" was the motto of the **W. H. Kiblinger Company.** The price was $250, in 1907, for a highwheeler, 2-passenger, 2-cylinder vehicle. The company was from Auburn, Indiana and built over 600 *Kiblinger* cars from 1907 to 1909.

_____

The **Zimmerman Manufacturing Company of Auburn** in Indiana was founded to build carriages and buggies. In 1909, they decided to enter the auto field with a 2-cylinder, 14-horsepower runabout that looked very much like a motorized buggy. In 1910, they added a 4-cylinder auto to their line. The Zimmermans, brothers Elias, Franklin, and John, then added a 6-cylinder auto in 1913. When Franklin and Elias both died, John closed the shop in 1914 and helped form the **Union Motor Company** to build the *Union* in 1915.

The *Union* of Auburn, Indiana was available in roadster and touring car styles with a 4-cylinder, 24-horsepower engine and sold for $895. The only year for *Union* production was 1916, when the brand went under for good.

1909 Zimmerman

From 1908 to 1909, the original name of the company that produced the *De Tamble* was

the **Speed Changing Pulley Co.,** of Indianapolis. The company changed their name to **De**

Tamble Motors Co. when they moved to Anderson, Indiana in 1909 until 1913. Their car was a

2-cylinder runabout and sold for $650. They offered 4-cylinder cars after 1909 and there were

four body types. The largest was the *Model K,* which was a seven-seater with a 120-inch

wheelbase.

The **Cloughley Motor Vehicle Co.** constructed their first steam car in 1896 but didn't

start producing them for sale until 1902. The Parsons, Kansas auto company used an 8 hp, 2-

cylinder, front-mounted engine in their vehicle. It was a rear-axle chain-driven, steam vehicle. It was offered with a gasoline engine in 1903, their last year of production.

The **Columbia Electric Company** started off making telephones and switched to autos in 1905, in McCordsville, Indiana. The car was called the *Leader,* and in 1905 was a 2-cylinder, 16-horsepower vehicle. In 1906 it was available in touring and runabout models.

In 1907, the company moved to more spacious quarters in Knightstown, Indiana. In 1909, with a change in ownership, 4-cylinder cars were added to the line-up and new models, including the *Demi-Torpedo,* were added. Upon the death of Luther Frost, one of the company principals, the company closed in 1912.

_____

Charles Rayfield of Springfield, Illinois is credited with inventing the carburetor, and his auto parts manufacturing business was thriving when he, with sons Bill and John, bought out the **Springfield Motor Car Company** in 1910. They used the factory to produce a 4-cylinder car in 1911 called the *Rayfield Junior Roadster* and a 6-cylinder version that was a 6-passenger roadster.

The company, now named the **Rayfield Motor Car Company,** relocated to a larger factory in Chrisman, Illinois in 1912. By 1913 the factory was producing over 200 vehicles a year. *Rayfield* 6-cylinder cars were water-cooled and had their radiators mounted in the rear of the vehicle. The hood was sloping, like the then-popular European styling of the *Renault.*

In 1914, the year that cyclecars were the most popular, **Rayfield** added what they called a cyclecar, although it was an atypical cyclecar with its 4-cylinder, water-cooled engine, rack-and-pinion steering, selective sliding-gear transmission, and a "pressed sheet" steel body.

In 1914 production was greater than 500 cars, and in 1916 it was over 600. Business was good until it was discovered that **Great Western Auto Company** of Peru, Indiana, who was contracted to assemble the cyclecar, went bankrupt and couldn't complete their contract. This also threw the **Rayfield Motor Car Co.** into receivership. In 1916, the company assets were auctioned for $14,000.

Birch Motor Cars, Inc. was a Chicago, Illinois company that sold their car, the *Birch,* by mail order from 1917 to 1923. A choice of 6-cylinder or 8-cylinder engines were offered, either the Herschell-Spillman, Lycoming, or Le Roi brands. Open and closed models were available.

Another mail-order car was the *Bush,* available from 1916 to 1924 from the Chicago-based **Bush Motor Company**. The 4- and 6-cylinder models used Lycoming and Continental engines.

A different car called the *Chicago* was built in 1915 and 1916 by the **Chicago Electric Motor Co.** of Chicago. Their electric car could seat five, go 23 mph, and could be operated from *either the front or the back seat*!

The **Duplex Motor Company** of Chicago manufactured the *Duplex* in 1908, so named for its double friction transmission and double driveshaft. The company was a subsidiary of the **Bendix Motor Company** of Logansport, Indiana, although all the *Duplexes* were built in Chicago. They had a style called the *Stanhope,* a *Surrey,* for their 1908 models, and a *Touring Roadster* in 1909, their last year. Their parent company, **Bendix**, went out of business shortly thereafter.

The **Crowdus Automobile Company** of Chicago, Illinois built a light electric runabout car called the *Crowdus Electric* from 1901 to 1903. It had a tubular frame and the steering tiller was used to accelerate and brake as well as steer. The vehicle's battery had a range of 50 miles per charge. Undercapitalization and poor sales caused the company to fail.

The **National Sewing Machine Company** of Belvedere, Illinois, manufactured the *Eldredge* from 1903 to 1906. With left-hand steering, the auto had its 8 hp engine located under the seat and it was connected to the rear axle by a chain drive and jackshaft. It's slogan, "Just what it ought to be" showed that it wasn't, when they went out of business in 1906.

The **Johnson Service Co.** of Milwaukee, Wisconsin manufactured a 4-cylinder, 50 hp steam car from 1905 to 1912. They also produced commercial vehicles.

Originated by Warren Johnson, who invented the electric thermostat in 1880, the vehicles were built with the money Johnson received for inventing the thermostat. The cars had 4-cylinder engines and were available in Runabout, Touring, and "Silent Comfort" models. Later models included Limo and Landaulet.

Johnson started out dabbling in commercial vehicles and had designed and built different one-ton steam truck styles, as well as a steam postal service wagon.

When Johnson died in December 1911, the company returned to making thermostats.

The **Monitor Automobile Works** of Chicago manufactured a 2-passenger high-wheeler called the *Monitor* in 1909. "Cars that run, stand up, and make good" was the motto for the car with the engine under the seat and a deck in the back.

In late 1910, they relocated the company to Janesville, Wisconsin, into a warehouse that was formerly used to store tobacco. A large 4-passenger surrey made its debut in 1911, which was converted into a *Milk Wagon/Pleasure Car*.

The company stopped making autos to make trucks exclusively, at the end of 1911. The motto for their trucks was "Designed Right, Priced Right, Built Right, All Right!" The trucks, called *Monitors,* were built through 1916.

———————————

In 1912, Walter A. Anger formed the **Anger Engineering Company.** In 1913, he assembled his first auto, the *AEC,* and it was available with both a 4-cylinder and 6-cylinder engine. Most of the vehicles would be custom-built and assembled upon being ordered, usually to the buyer's specifications. The autos were made until 1915. Despite the name, there were no reports of road rage associated with the brand.

The **Earl Motor Co.** (not to be confused with the one in Jackson, Michigan) was a short-lived marque of Kenosha, Wisconsin. Although they claimed manufacture of a few cars in Milwaukee, it is widely believed their first car was manufactured in 1907 in Kenosha, in the former **Visible Typewriter Company** factory.

The company failed in 1908, and the **Petrel Motor Car Company** moved in.

———————————

The **Petrel Motor Car Company** consisted of Samuel Watkins, the president of the **Beaver Manufacturing Co.**, an engine-making concern, and brothers John and Harry Waite, who had built a few cars in the back of a rented store in Milwaukee.

In 1909 their car with a friction transmission and double chain-drive was at first offered in 6- or 4-cylinders, but the 6-cylinder version was discontinued after the first year. Thereafter, the car's motto was "The Aristocrat of Medium-Priced Cars."

In late 1909 they moved out of the former typewriter factory and into a factory on Virginia Street in Milwaukee.

By 1911 the company was in trouble and Sam Watkins sold it, along with the **Beaver Manufacturing Co.,** to Corliss steam engine builders **Filer and Stowell.** They moved the enterprise to the outskirts of Milwaukee and continued to produce the *Petrel* and a new vehicle called the *F. S.* They were out of business by the end of 1912, with about a total of about 1,000 *Petrels* and *F. S.* cars built.

———————————

In 1901, Joseph Merkel bought the Milwaukee factory of the **Layton Park Manufacturing Co**. to use for his **Merkel Manufacturing Co**. In the first years he built motors, bicycles, and the *Flying Merkel* motorcycle.

In 1905 he manufactured the *Merkel,* a small, 4-cylinder car available with water-cooled or air-cooled engines. He called his models "forms" and had three available. The cars had shaft-drive and prices ranging from $1,500 to $3,500.

When the manufacture of the auto wasn't profitable, Merkel returned to the more stable production of motors, bicycles, and the *Flying Merkel.*

––––––––––––

The *Excalibur* was built by the **Beassie Engineering Company** in Milwaukee in 1952 and 1953. It was designed by Brook Stevens originally to resemble the 1928 *Mercedes SS* and has gone through many changes. The *Excalibur* auto would go in and out of production and business over the decades.

A new prototype premiered at the 1964 New York Auto Show which was developed in conjunction with the **Studebaker Motor Company**. It had a Studebaker *Lark Daytona* chassis and a Studebaker 290 horsepower V-8 engine. Even though Studebaker went out of business in 1963, this didn't stop the *Excalibur*. Stevens went to General Motors and secured new engines for the cars. He then formed his own company to sell them. From 1965 to 1969 the company produced the Series 1 *Excalibur*.

The new vehicles had excellent acceleration and could go from 1 to 60 mph in less than

6 seconds and had a projected top speed of 134 mph. Over 3,200 were manufactured in this second incarnation of the company.

The next time the *Excalibur* was manufactured was from 1970 to 1974. By this time, Brook Stevens was joined by his sons Steve and David. The Series II had models from $6,000 to $13,000, with air conditioning, positraction, power steering, chrome wire wheels, stereo radio, adjustable steering wheel, and more. Speeds of 149 mph were advertised.

Series III was manufactured from 1975 to 1979 and featured 1,141 cars produced. Series IV was from 1980 until 1984, when 994 autos were manufactured. The Series V was from 1985 to 1989 and 389 autos were produced. The *Limited Edition 100* and the *Excalibur Cobra* were built from 1990 to 1994. A famous *Excalibur* promoter was Phyllis Diller.

In 1906, 17-year old Harry House got some press attention from the Cheyenne, Wyoming newspapers. He was working on a car with a motorcycle engine and various bicycle parts. Nothing more was heard after 1906.

Vincent Bendix bought the **Triumph Motor Car Company**'s factory in 1908 and produced his car the *Bendix* through his **Bendix Company** (as well as continuing to produce the *Triumph* for a short time.) He moved the factory to Logansport, Indiana and his **Duplex Motor Car Company** manufactured autos in Chicago. By 1909, the company was out of business. Bendix, through the **Bendix Company,** went on to manufacture auto parts, including braking systems, magnetos, and generators, for autos and aircraft.

Destroyer of Auto Companies—Archie Andrews and The *Ruxton*

One car had its prototype built in New York, and then was built in St. Louis, Missouri before moving to Kenosha, Wisconsin. It was the *Ruxton*.

Most people, when they hear the name "Archie Andrews," they think of a red-headed teenager from the comic books, with a friend named Jughead and who can't decide between Betty and Veronica, who all play in a band named for Archie. However, a real-life Archie Andrews might be even more deserving of his own comic book, if only as the arch-villain. For how many men could kill not one, not two, but FOUR different auto companies?

Andrews formed his own company, **New Era Motors**, to produce the car. Since New Era didn't have production facilities, Andrews manipulated stock to take over the Moon Motor Co. Familiar with Andrews, the previous directors of the Moon Motor Co. BARRICADED THEMSELVES IN THE OFFICE, with their armed Pinkerton security guards, and refused to let Andrews in! Andrews eventually gained entrance with the help of a court order and the St. Louis Police Department, who rushed the office with guns drawn.

The *Hupmobile* was a popular car for many years until Archie Andrews came along. Andrews became enamored of a car he saw called the Ruxton. It was named for William Ruxton, a financier who it was hoped would invest in the car. He didn't, and later sued to have his name removed from the car.

The Ruxton

The *Ruxton* was considered a marvel of engineering, a front-wheel drive car designed by engineers from the Wills Ste. Claire plant with C. Harold Wills himself taking part. Andrews took the design and insisted that **Hupp Motor Company** manufacture the car. It was a very different model than the ones **Hupp** was then producing and retooling the factories to produce the car would have been expensive, so the board refused. To mollify him, a prototype was produced but the car wasn't produced.

Unfortunately, the facilities of the **Moon Motor Co**. were inadequate to build the *Ruxton*. Besides the **Moon Motor Co**., Andrews had interest in the **Kissel Motor Car Company**. Besides producing their own auto, the **Kissel Motor Car Co**. also made parts for other companies, including **Moon Motor Company**. They were commissioned to make the transmissions and final-drive assemblies for the Ruxton. Andrews moved production to the Kissel factory.

The Kissel brothers at first put up with Andrew's intrusions, but when it became evident he was going to try to take over their company too, the brothers voluntarily took the company into receivership! In other words, they decided they'd rather close up shop than let the company fall into Andrew's hands.

With the **Kissel Motor Car Co**. out of business, this left the **Moon Motor Company** at a disadvantage because Kissel had been one of their major suppliers. Soon, after only producing about 525 Ruxtons, the **Moon Motor Company'**s resources were totally depleted, and they filed for bankruptcy in November of 1930, as did **New Era Motors** five days later.

With **Kissel, Moon**, and **New Era** gone, Andrews turned his attention back to the **Hupp Motor Car Company** where he was still on the board. In 1934, he hustled his way into the chairmanship. By 1935, the old board and the courts forced him back out but not soon enough. **Hupp Motor Car Co**. was weakened through its board battles enough that it produced no cars in 1936. By 1938, the company was weakened enough by Andrews that it went into receivership.

Ok, for those keeping score, here is the list of car companies ruined by Archie Andrews:

1. Moon Motor Company

2. Kissel Motor Car Company

3. New Era Motors

4. Hupp Motor Car Company

In 1936, Andrews lost his seat on the Chicago Stock Exchange due to improprieties. In 1937, his properties were placed in federal receivership. Archie Andrews died in 1938 at the age of 59. According to some accounts, he died at his Greenwich, Connecticut mansion. According to others, he was facing a jail sentence and died on the run-in Canada!

Archie Andrews, Hupmobile Sales Convention

George Dorris and John French started the **St. Louis Carriage Car Company** of St. Louis, Missouri, in 1899 and were the first successful car company (or so they said) west of the Mississippi. They built a factory on North Vandewater St. in St. Louis, said to be the first factory built expressly for auto production. Over 600 cars were manufactured from 1899 to 1905.

The cars were 1- and 2-cylinder models, with a chain drive and steered with a tiller. One of their mottos was, "Rigs that Run." In 1902 a steering wheel was introduced, as well as a 4-cylinder model.

Although the company was enjoying a good measure of success in St. Louis, French decided to move the operation to Peoria, Illinois in 1905, where they moved into a new, large, 3-story factory located on 10 acres of land. Unfortuneatly, it was too much too soon, and the company went under. The Peoria factory was sold in December of 1907 for $10,000.

The *Lexington* was first built in Lexington, Kentucky, by the **Lexington Motor Company** in 1909. The next year it moved to Connersville, Indiana, and stayed until they went out of business in 1927. A Connersville businessman, E.W. Anstead, who manufactured axles and springs, bought the company in 1912 and produced it alongside another car, the *Howard,* until 1914.

Six-cylinder cars began to be produced, and some of the more popular were the *Thoroughbred Six* and the *Minute Man Six.* The cars were available in many models, including, besides the usual touring and runabout styles, models called *"Salon Sedan," "Sedanette," "Lark Sport Touring," "Royal Coach," and "California."*

Over 40,000 *Lexingtons* were made before the company ended in 1927. The whole enterprise was taken over by the **Auburn Motor Company**, which phased out the *Lexington.*

———————

Frank H. Morse built a 4-wheel drive steam auto in 1902, in his native Milwaukee, Wisconsin. He sold it for $550. The next year he joined the **Four-Wheel Drive Wagon Company** in Milwaukee, working on manufacturing commercial vehicles. In 1907, he started the **Wisconsin Motor Manufacturing Company,** which he operated until 1909. Then he joined the **Kissel Motor Company** until 1912, when he relocated to Pittsburgh, worked on the *Duquesne*, and then went to work on his own cyclecar.

The *Morse* was a front-wheel drive auto, when most cars of the day were rear-wheel drive. Developed and manufactured in Pittsburgh, PA from 1914 until 1916, it had a 2-cylinder, 9-horsepower engine. It looked like a winner, but the new **Morse Cyclecar Company** was out of business in a year.

The **Downing Cyclecar Car Company of Detroit** started in Detroit in 1914 and then built

a factory and moved many operations to Cleveland, Ohio in 1915. More of a light car than a

cyclecar, it sold for $500 and had a 4-cylinder model as well as the 2-cylinder ones. After

moving to Cleveland, the *Downings* that had been built in Detroit were called the *Downing-*

Detroit models. After the 1915 models, the company went out-of-business.

_____

The **Johnson Service Co.** of Milwaukee, Wisconsin manufactured a 4-cylinder, 50 hp

steam car from 1905 to 1912. They also produced commercial vehicles.

Originated by Warren Johnson, who invented the electric thermostat in 1880, the

vehicles were built with the money Johnson received for inventing the thermostat. The cars had

4-cylinder engines and were available in Runabout, Touring, and "Silent Comfort" models. Later

models included Limo and Landaulet.

Johnson started out dabbling in commercial vehicles and had designed and built

different one-ton steam truck styles, as well as a steam postal service wagon.

When Johnson died in December 1911, the company returned to making thermostats.

In 1912, Walter A. Anger of Wisconsin formed the **Anger Engineering Company.** In 1913,

he assembled his first auto, the *AEC,* and it was available with both a 4-cylinder and 6-cylinder

engine. Most of the vehicles would be custom-built and assembled upon being ordered, usually to the buyer's specifications. The autos were made until 1915. In spite of the name, there were no reports of road rage associated with the brand.

The **Earl Motor Co.** (not to be confused with the one in Jackson, Michigan) was a short-lived marque of Kenosha, Wisconsin. Although they claimed manufacture of a few cars in Milwaukee, it is widely believed their first car was manufactured in 1907 in Kenosha, in the former **Visible Typewriter Company** factory.

The company failed in 1908, and the **Petrel Motor Car Company** moved in.

_____

The **Petrel Motor Car Company** consisted of Samuel Watkins, the president of the **Beaver Manufacturing Co.**, an engine-making concern, and brothers John and Harry Waite, who had built a few cars in the back of a rented store in Milwaukee.

In 1909 their car with a friction transmission and double chain-drive was at first offered in 6- or 4-cylinders, but the 6-cylinder version was discontinued after the first year. Thereafter, the car's motto was "The Aristocrat of Medium-Priced Cars."

In late 1909 they moved out of the former typewriter factory and into a factory on Virginia Street in Milwaukee.

By 1911 the company was in trouble and Sam Watkins sold it, along with the **Beaver Manufacturing Co.,** to Corliss steam engine builders **Filer and Stowell.** They moved the enterprise to the outskirts of Milwaukee and continued to produce the _Petrel_ and a new vehicle called the *F. S.* They were out of business by the end of 1912, with about a total of about 1,000 *Petrels* and *F. S.* cars built.

In 1901, Joseph Merkel bought the factory of the **Layton Park Manufacturing Co**. to use for his **Merkel Manufacturing Co**. In the first years he built motors, bicycles, and the *Flying Merkel* motorcycle in Wisconsin.

In 1905 he manufactured the *Merkel,* a small, 4-cylinder car available with water-cooled or air-cooled engines. He called his models "forms" and had three available. The cars had shaft-drive and prices ranging from $1,500 to $3,500.

When the manufacture of the auto wasn't profitable, Merkel returned to the more stable production of motors, bicycles, and the *Flying Merkel*.

The *Excalibur* was built by the **Beassie Engineering Company** in Milwaukee in 1952 and 1953. It was designed by Brook Stevens originally to resemble the 1928 *Mercedes SS* and has gone through many changes. The *Excalibur* auto would go in and out of production and business over the decades.

A new prototype premiered at the 1964 New York Auto Show which was developed in conjunction with the **Studebaker Motor Company.** It had a Studebaker *Lark Daytona* chassis and a Studebaker 290 horsepower V-8 engine. Even though Studebaker went out of business in 1963, this didn't stop the *Excalibur*. Stevens went to General Motors and secured new engines for the cars. He then formed his own company to sell them. From 1965 to 1969 the company produced the Series 1 *Excalibur.*

The new vehicles had excellent acceleration and could go from 1 to 60 mph in less than 6 seconds and had a projected top speed of 134 mph. Over 3,200 were manufactured in this second incarnation of the company.

The next time the *Excalibur* was manufactured was from 1970 to 1974. By this time, Brook Stevens was joined by his sons Steve and David. The Series II had models from $6,000 to $13,000, with air conditioning, positraction, power steering, chrome wire wheels, stereo radio, adjustable steering wheel, and more. Speeds of 149 mph were advertised.

Series III was manufactured from 1975 to 1979 and featured 1,141 cars produced. Series IV was from 1980 until 1984, when 994 autos were manufactured. The Series V was from 1985 to 1989 and 389 autos were produced. The *Limited Edition 100* and the *Excalibur Cobra* were built from 1990 to 1994. A famous *Excalibur* promoter was Phyllis Diller.

Vincent Bendix bought the **Triumph Motor Car Company**'s factory in 1908 and produced his car the *Bendix* through his **Bendix Company** (as well as continuing to produce the *Triumph* for a short time.) He moved the factory to Logansport, Indiana and his **Duplex Motor Car Company** manufactured autos in Chicago. By 1909, the company was out of business. Bendix, through the **Bendix Company,** went on to manufacture auto parts, including braking systems, magnetos, and generators, for autos and aircraft.

From 1906 until 1909, the **Reliable-Dayton Motor Car Company** built autos in their Chicago, Illinois factory. The car was a high-wheeler, with rope drive, solid rubber tires, and a 2-cylinder engine. Under the hood were the gasoline and water tanks. Under the seat was the engine. In 1909, the factory was taken over by the **FAL Automobile Company.**

In 1906, Willis Copeland of the **Single Center Buggy Co.** brought designer and engineer William O. Worth to Evansville, Indiana to help build an automobile that he could sell. Copeland had previously struck out with the *Zent.*

Discord began when Worth designed a different car than Copeland had ordered. Worth's was a high-wheeler design, with a 2-cylinder, air-cooled engine. It had double chain drive, a friction transmission, and was steered by a tiller.

William Worth revolted and formed **Worth Motor Car Manufacturing Company**. After making a few cars in Evansville in 1907, in 1908 Worth moved to a Kankakee, Illinois factory. Unfortunately, he was evicted for not paying the rent. He found another place to assemble his autos—an old cattle pavilion. He struggled here until October 1910 when he declared bankruptcy. His vice-president left with the last car in the factory, stuffed with as many auto parts that would fit in it.

In 1909, Frank Enger of Cincinnati formed the **Enger Motor Car Co.** to manufacture his 2-cylinder, high-wheeler automobile. In 1910, he switched to a 4-cylinder, overhead valve engine of his own design. Production was continuous, with about 300 to 400 cars produced each year.

In 1915, Enger offered a six-cylinder of his own design, but the best was yet to come! In late 1915 he introduced the Enger Twin Six engine, which was a 12-cylinder engine, one of the first in the business. In 1916 he developed a way for only six of the cylinders of the 12 to be used.

At the end of 1917, a successful auto operation ended when Fred Enger was diagnosed with cancer and shot himself to death in his office. Although leaving instructions to his Vice-President about carrying on, his widow wanted no more of the auto business and had the company liquidated. With World War I on the horizon, the factory was re-tooled for war work.

The Enger 1917 Six-Twelve Touring Model.

The *Lanpher* was a high-wheeler vehicle manufactured in 1906 to 1916 in Carthage, Missouri. It had a planetary transmission, double-chain drive, with a 2-cylinder, air-cooled engine. It sold for $550 and was invented by brothers Earl and Norman Lanpher, proprietors of **Lanpher Brothers Carriage Works.** The carriages were built right alongside the cars, with Earl taking charge of the carriage end, and Norman watching the autos.

The business remained chiefly a local concern, with most of the advertising word-of-mouth. In 1916, the brothers left Carthage, ending the business.

—————————

In 1908, the **Economy Motor Buggy Company** of Fort Wayne, Indiana, built a 2-cylinder high-wheeler that the **Success Motor Company** of St. Louis, Missouri said infringed on their patent. Upon not receiving a royalty, they sued the **Economy Motor Buggy Company.** Instead of defending the suit, the **Economy Motor Buggy Company** skipped town and quietly moved to Kankakee, Illinois, and then started car production again in a Joliet, Illinois factory. In 1910 and 1911, they concentrated on making commercial vehicles.

The **Success Motor Company** eventually found them in Joliet, and after their lawsuit the Economy Motor Buggy Company declared bankruptcy in 1911. William Pratt of **Pratt Manufacturing Company** purchased the remains. William Pratt was involved with the **Pratt-Elkhart Motor Company**, starting in 1909.

—————————

In 1905, in a small workshop on Front Street in Woodburn, Indiana, the **Woodburn Auto Co.** was born. The vehicles were of the high-wheeler style, with 2-cylinder, air-cooled engines.

They also made a few trucks and tractors. The cars were sold locally, and about 12 were put together in all. In 1912 they shuttered their doors.

———————

In 1909, the **Ricketts Automobile Company** manufactured their first car in South Bend, Indiana. Thomas and then Joseph Ricketts were the heads of the company. Joseph was convinced the early low sales were due to the negative connotation of the name and the disease caused by a Vitamin D deficiency called "rickets."

He at first changed the name to the **Diamond Motor Company,** and then **R.A.C. Automobile Company,** the initials of the original company name. The vehicle was still using model names, including the *Model F, Model H, Model D, and the Model G-6.*

When the name was changed to **R.A.C.**, the motto was "New in name, and practically a new vehicle." But it was too little, too late as the company went into receivership in 1911.

The Henderson Motor Company of Indianapolis manufactured a two-seater and 5-seater autos, with 4-cylinder engines and *Stutz* transmissions from 1912 until 1915. The sons of the founder went on to invent the *Henderson* motorcycle.

The **Black Manufacturing Company** at 216 West Berry Street in Fort Wayne, Indiana, shared space with Alfred Randall's car dealership. In 1906, Randall tried his hand at auto manufacturing and assembled a 2-cylinder, 12 horsepower runabout, naming it for Marion

Black. Selling for $750, by 1909 the company was downsized into only selling auto parts. In 1916, Randall sold his auto dealership and went into auto financing.

The **Reeves Pulley Co.** of Columbus, Indiana, was the main business for Milton O. Reeves. Cars were at first just a sideline when he built his first autos from 1896 to 1898 and then again from 1905 to 1912. His first car, the *Motocar,* was so loud he developed one of the first mufflers to deaden the sound. With a double-chain drive and an air-cooled Sintz engine, he sold five.

In 1898, an announcement was made that the **Reeves Pulley Co**. would not be manufacturing any more autos so that they could concentrate on their engines and the variable speed transmission (VST) they had developed. In 1905, they re-entered the car business as they built a motorcar with their new, 4-cylinder engine.

The new engine had found favor with coal baron Alexander Malcomson, who ordered 500 of the engines for his soon-to-come *Aerocar.* When the *Aerocar* went out of business just a year after it started, Reeves decided to build his own cars to put the engines in. He produced several 4- and 6-cylinder vehicles.

In 1907, Reeves came out with the *Go-Buggy,* a high-wheeler with a 2-cylinder air-cooled engine and a double-chain drive. It sold for $450, and you could pay for additional coachwork. This model proved popular with Indiana farmers, but in 1910 Reeves decided to once again concentrate on his lucrative industrial pulley system.

In 1911, without manufacturing more than a prototype, Reeves decided to test the waters a bit by taking to the New York Auto Show an experimental vehicle he had been working on—the 8-wheeled *Octoauto*. When this car failed to light any fires, he tried again, with the same results, with a 6-wheeled vehicle called the *Sexto Auto*. When these cars failed, Reeves returned to his other business interests.

The **Rider-Lewis Motor Co.** operated in Muncie, Indiana from 1908 to 1910. They advertised "The Excellent Six," their overhead camshaft and valves car engine with 50 hp. They also produced a 4-cylinder, 26 hp model. The body types included a two-seater roadster, a 5-seater touring car, and limousines.

For the year of 1898, the *Eichstaedt* auto was built at 121 West Market St. in Michigan City, Indiana. The inventor was Roman Eichstaedt, who was a pattern-maker and worked for the railroads and local machine shops before leaving to start his own company. His auto was a two-seater, one-cylinder vehicle that could attain speeds of 20 miles per hour.

Although he still built the occasional car on request until 1902, he quit auto-making to concentrate on his bicycle, the *Roman Bicycle,* which proved to be more lucrative for him.

Mechanic Schuyler Zent of Marion, Ohio, built the *Zentmobile* in 1903. The one-cylinder, water-cooled two-seater was built in the factory of Willis Copelands' **Single Center Buggy Company** in Evansville, Indiana. With Copeland in charge of manufacturing the auto, Zent's job was to sell it. Something didn't gel, and the partnership only lasted a year. Zent moved back to Ohio.

Logansport, Indiana had the **Motor Car Corporation** from 1917 to 1926. They produced the *Revere,* which was a luxury car with sportster aspects. With a 4-speed transmission and 103

 hp engine, it was powered by a Rochester-Duesenberg engine. Popular among auto aficionados for a few years, it fell out of favor and out of business

----

Willis Copeland decided to form the **Evansville Motor Co.** in Evansville, Indiana to build the auto called the *Simplicity* from 1906 until 1911. Offering roadsters, touring cars, and limousines, the various styles had 4-cylinder, water-cooled engines. A friction-transmission was used, which did away with the clunky chains. The transmission proved popular and worked well on a nice day but was faulty when it rained.

Water would leak into the friction gears and the car would start but wouldn't move. Once the engine was enclosed, the problem was solved. But the inventor Willis Copeland felt that the car's reputation had been besmirched and changed the company name to **Traveler Automobile Company** in 1909 but continued to manufacture the *Simplicity*.

By 1912, Copeland had tired of producing automobiles and confined himself to servicing the cars he had already made. He later became a dealer for the *Flint* auto and the *Chevrolet*.

The **Great Western Automobile Company** was a bicycle establishment when they experimented with building a steam car in 1902. They were still experimenting in 1905, without actually manufacturing a car.

Since the first **Great Western Automobile Company** never actually manufactured a car, a new company from Peru, Indiana was free to use the name in 1910.

Their first car was a touring car, with 7 seats and a 50 hp engine. They manufactured a smaller touring car also, with a 30 hp engine. They manufactured their cars for eight years, from 1908 until 1916.

In 1910, they released a 4-cylinder, 40 hp engine in a standard chassis with a 114-inch wheelbase called the *Great Western 40*. It was their most popular model, available in five different styles.

Also, from Peru, Indiana was the **Bryan Steam Motors Co.,** which from 1918 to 1923, manufactured one car per year, for a grand total of six cars. The autos were powered by steam and weighed about 4,500 pounds.

The **Single-Center Buggy Company** wasn't a mobile dating club but a company in Evansville, Indiana that made the *Single-Center* auto. It had a 2-cylinder, water-cooled engine and was sold from 1906 to 1908. With a wheelbase of only 84 inches, it was available with pneumatic or solid rubber tires.

In Mason City, Iowa, William Colby established the **Colby Motor Company** in 1911, and produced 40 hp four-cylinder vehicles—later a smaller 30 hp was offered.

In December 1911, the company merged with the **National Cooperative Farm Machinery Company of Davenport.** In late 1913, the company was purchased by the **Standard Motor Company** and even later, was purportedly kept alive by a rich Iowa widow who would come to the company's aid with large checks whenever summoned!

When the company breathed its last in 1914, its total production had been a total of 561 vehicles.

Leslie Hobbie at first followed his father into the blacksmith business but became a car enthusiast and started a dealership in Hampton, Iowa. The yearning to create his own car caused him to assemble and introduce the *Hobbie Accessible* auto in 1908. The car was a high-wheeler style, able to be steered with a tiller, with solid tires. It had twin cylinders, an air-cooled engine, solid tires, and used a tiller. The models offered were "Road Wagon" and "Piano Box Buggy." In 1910 a 2-passenger Runabout model was added.

By 1911, the **Hobbie Automobile Company** was once again an auto dealership. Hobbie sold *REO, Ford, Premier, Olds,* and later *Overland* autos.

The **Adams Automobile Company** of Hiawatha, Kansas built the strangely named *Average Man's Runabout.* Selling for $500, the 2-cylinder, 10 horsepower vehicle was only sold for one year, 1906. George Adams, the founder, built cars for personal use only in 1907 and 1908.

In 1903, the **Bates Motor Company** of Lansing, Michigan, had a motto- "Buy a *Bates* and Keep Your Dates." But not enough people took this advice and by 1905, the 4-seater, 20 hp, 3-cylinder touring model car with 3 forward speeds, was defunct. The $2,000 price was a big factor in the company's demise.

The **Haroun Motor Sales Corporation** of Wayne, Michigan, not far from Detroit, had a respectable career. They manufactured the *Haroun, an* open model auto, from 1917 until 1922. The cars were named after Ray Haroun, the winner of the first Indianapolis 500. The company built their own 4-cylinder engines for the car.

In 1916, engineer R.C. Aland raised a half-million dollars to launch the **Aland Motor Car Company** in Detroit. The car they built, the *Aland*, had, according to the advertising, "a racing-type aluminum engine." The engine was a 4-cylinder, with each cylinder having a single overhead camshaft and valve. It was rated at 14 horsepower, and because of its efficiency, Aland claimed it could exceed 65 mph.

The company offered a five-passenger touring model selling for $1500, and a less-expensive 2-passenger roadster. The car had diagonally-connected internal expanding brakes, on all four wheels! This was unusual for the times, but not enough to save the company. Their last year was 1917.

1917 Aland 5-Passenger Touring Car

Another unusual Detroit car name was the *Spider,* to be manufactured by the **Detroit Body Company.** Another of the many cyclecar attempts produced in Detroit in 1914, the company was located at Clay Street and St. Aubin Street in Detroit. They apparently never made it beyond the prototype stage.

The *Hammer* was manufactured by the **Hammer-Sommer Auto Carriage Co., Ltd.** in Detroit, Michigan. It was small, with an 82 inch wheelbase, with a 2-cylinder. 12 hp gasoline engine and called the *Hammer-Sommer* in 1904 and 1905.

The partners, Henry Hammer and Herman and William Sommer, quarreled and went their separate ways in 1906. Hammer produced an auto that was double the Hammer-Sommer—it had a 4-cylinder, 24-hp engine, with a 100-inch wheelbase. It was called the *Hammer.* The Sommer brothers continued to manufacture the *Hammer-Sommer* as just the *Sommer.* By 1906, both the **Hammer Motor Company** and the **Sommer Motor Company** were out of business.

In 1903, the **Blood Brothers**, Maurice and Charles, and the Fuller Brothers, Charles and Frank, combined forces to form the **Michigan Automobile Company, Ltd.** The Bloods were mechanics who owned the **Kalamazoo Cycle Company** and the Fullers were businessmen who owned a factory that manufactured wooden devices.

1903 Michigan Runabout

The first car they constructed in 1903 was deemed too small, with a 48-inch wheelbase. The next try was called the 1904 *Michigan* and had a 54-inch wheelbase. In 1905 the Bloods and the Fullers quarreled and parted company, with the Bloods leaving the **Michigan Automobile Co.** and forming the **Blood Brothers Automobile and Machinery Co.** They continued producing the same auto but named it the *Blood* instead. The Bloods were concurrently agents for the *Pope* and *Cadillac* cars. After 1906, the *Blood* was not produced any more. A year later, the Fuller Brothers stopped producing the *Michigan*.

The **Blood Brothers** were back in 1914 with a cyclecar called the *Cornelian*. It was first manufactured in Kalamazoo and in 1915 moved production to larger quarters in Allegan, Michigan. The vehicle had a full-floating rear axle and independent rear suspension, revolutionary advancements for the time.

The 1915 *Cornelian* entered the Indianapolis 500 and became the smallest car ever in the race. The driver was Louis Chevrolet, co-founder with William Durant of **Chevrolet Motor Company**. The car made a good showing and sales grew. In June of 1915 the **Blood** factory was employing 248 men, with shifts going 24 hours ostensibly to produce 25 *Cornelians* per week. But by October the bottom had fallen out of the cyclecar market. A little over 100 *Cornelians* had been made when it was announced that the *Cornelian* would be discontinued.

David Buick was responsible for the familiar white porcelain bathtubs seen everywhere. He was the inventor of the process to affix porcelain enamel onto cast-iron, still used for the modern bathtub. Another invention of his was a lawn sprinkler—he had thirteen patents in all when he got together with his son Tom in 1900 and designed an engine and built his first car.

Buick became fixated on designing autos and sold his plumbing business to finance his auto obsession. He took in other investors, the Briscoe Brothers to help finance his project. But when Buick took over a year to produce his prototype, the Briscoes sold out to the **Flint Wagon Works** and started backing Jonathan Maxwell, whose **Maxwell Auto Company** went on to be the forerunner of **Chrysler Auto Works**.

The Buicks ended up owning only 3% of the company and sold out to Billy Durant, who went on to purchase controlling interest in Buick and use it to start **General Motors.**

In the meantime, David Buick was completely out of the company that bore his name by 1908. He moved to California and organized an oil company. He lost the rest of his money in that venture due to legal problems over land ownership.

In two years, he was back in Michigan and tried his hand at starting another motor company, The **Lorraine Motor Company**. When that one failed, he tried again with the **David Dunbar Buick Company**, putting out a car called the *Dunbar*. When this company also failed, he went to Florida, which was having a land boom with lots of people making money. Buick didn't though, when the real estate firm he was with went out of business.

He finally ended up as an instructor at the Detroit School of Trades, until he was demoted to working the night information desk in 1927. He died penniless. Friends chipped in to buy him a tombstone that carried the Buick logo.

The Buick 6 was the first renowned and popular car for General Motors.

1911 Buick

1932 Buick

The Briscoe Brothers, Frank and Benjamin, helped to finance David Buick's first cars. Buick kept going back to the Briscoe Brothers for more money until the Briscoes owned 97% of the company. When the Briscoes met Jonathan Maxwell, they got tired of David Buick's tinkering-they didn't even have a prototype auto yet! They sold their interest in **Buick** and invested in the *Maxwell* auto, which had a fair amount of success.

The company that they sold their **Buick** shares to, the **Flint Wagon Works**, eventually fell into the control of Billy Durant, and when Buick's car eventually saw the light of day, it became the premier car of **General Motors** auto company.

The **Huber Automobile Company** was organized in Detroit in 1903 with $100,000. The company had originally intended to manufacture automobiles but got sidetracked into making tour buses. Soon, the company was more involved in renting and leasing their tour buses than continuing any further manufacturing. They went into liquidation in 1907.

Flint, Michigan was known as the Vehicle City, not for being the birthplace of **General Motors**, but because three major carriage makers were in Flint. One was the **Durant-Dort Carriage Co.**, one was the **W.A. Paterson Co.**, and the third was the **Flint Wagon Works.** All three ventured into auto production.

The **Flint Wagon Works** was the company that first purchased the rights to the *Buick* from David Dunbar Buick. The **Flint Wagon Works** sold the rights to Billy Durant, who used the company to start **GM**.

The **Flint Wagon Works** then formed the **Whiting Motor Car Company** in 1910 and manufactured the *Whiting,* a 4-cylinder, shaft-driven, medium-sized auto, with sliding-gear transmission. The auto was designed by James Whiting.

The car proved to be so successful that William Durant bought the **Flint Wagon Works** to get it. Then he shut down the marque in 1912 and never made any more *Whitings.* He used the factory of the **Flint Wagon Works** to make *Chevrolets*.

1910 Whiting

―――――――

William Durant's *Star* had a 4-cylinder engine and only cost $443, priced to compete with Ford's Model T. The *Star* was part of **Durant Motors** conglomerate and did well—in 1923 it was the #7 selling car in the nation. But overall problems with the main conglomerate caused the *Star* to cease production in favor of the *Maxwell,* the strongest car and company in the **Durant Motors** line-up.

1907 Maxwell Briscoe Roadster

————————————

During the cyclecar craze, from 1914 to 1915 the *Malcolm* was manufactured by the

Malcolm Jones Cyclecar Co. in Detroit. Despite being the typical cyclecar of 2 cylinders, the

Malcolm had a 4-cylinder "en bloc" 18-horsepower engine.

It was a small car but did have room for three passengers. A V-twin engine, friction

transmission, belt-driven model called the *Malcolm Jones* was also offered.

————————————

The 1895 *Ames* was a "buggy box and seat suspended between two bicycle frames." It was propelled by a boiler providing steam to each of the pistons on each bicycle frame. It was built in Owatonna, Minnesota by A.C. Ames and the **Ames Motor Cycle Company.**

In Mankato, Minnesota, candy-maker Ernest Rosenberger put together a 4-wheel-drive vehicle in 1909. He felt 4-wheel drive would help navigate the hilly terrain. Forming the **Four-Wheel Traction Co**. to manufacture the car, the town held a ceremony opening the new factory in 1907. There were three workers in all, and they had assembled five cars by 1908.

By 1912, over 30 cars and trucks had been made. The factory continued building *Katos,* as the cars were called, until 1913, when the **Four-Wheel Traction Co.** sold out to the **Nevada Manufacturing Co.** of Nevada, Iowa.

Olaus Lende built the *Lende* in Granite Falls, Minnesota, from 1902 to 1909. After moving to Granite Falls around 1898, Olaus endeared himself to the community right away as he wired the town for electricity, long before any of the neighboring communities had it. He became interested in autos in 1902 while working in John Iverson's foundry. A customer asked him to look at a broken axle on his auto.

LENDE

After getting started working on cars, Lende built his own 2-cylinder, gasoline-powered vehicle. He soon sold it and built two more, which he also sold. He developed a 4-cylinder, air-cooled engine with a single-coil ignition system for his next car. It was ahead of its time, as was its use of a generator to charge a storage battery for the car's ignition and lights. The new autos sold as fast as Lende could make them, especially after the cars won a few local races.

With plans to produce more autos, Lende formed the **Lende Automobile Manufacturing Company** and opened an office in Minneapolis in 1908. The planned 1909 vehicle received some national press when it was announced that the company would release an auto with 4-cylinders and 30 horsepower, shaft drive, and planetary transmission, selling for $1,800. But the venture died out by late 1909, after only 17 autos were manufactured—most of them custom-built. The farthest away a *Lende* was sold was to a customer in Watertown, South Dakota.

After his car-manufacturing career, Otis Lende continued as an auto mechanic and dealer. In his dealership, he sold *Star*, *Saxon*, and *Studebaker* automobiles.

_____

The **Dispatch Motor Company** of Minneapolis, MN built the *Dispatch* in 1910. It was a 2-cylinder car. At first, they were a mystery company, since their first ad in the *Cycle and Car Trade Journal* of December 1910 apologized for being back-logged on filling the 1910 orders but promising to have them done soon. The thing is, this was the first anyone had heard of the company, so how could there be a back-log?

The ad then went on to describe its upcoming 4-cylinder autos of 1912. Curiously, the company kept updating their information about their upcoming autos apparently without actually producing any. Although the company was listed in business directories until 1913, and the Minneapolis City Directory until 1923, any manufacture beyond 1910 is doubtful. It is surmised to have been a garage or dealership in the following years. It is ventured that they only built one car, the original 2-cylinder one, but whether the information leaked to the press was an intended stock scam, or an actual attempt to get started in auto manufacturing, has never been exactly revealed.

The **Beggs Motor Car Co.** of Kansas City, Missouri built a car with a 6-cylinder, Continental motor from 1918 until 1923. They started off making buggies and specialized in ornate, decorative circus wagons.

In July 1921, disgruntled stockholders attempted court proceedings to take the company into receivership. The judge disagreed, and the company stayed together until 1923, when it was once again sued, and this time dragged into receivership. The litigation didn't end until 1926.

The *Darby* was named for designer C.T. Darby and financed in 1909 by two St. Louis auto enthusiasts who called their company the **Darby Motor Car Company**. Advertised as the "Simplest Automobile on Earth," the *Darby* had 2-cylinders, 16-horsepower, and a 100-inch wheelbase, available in a 3-passenger roadster or a 4-passenger surrey model. Its engine was a two-stroke and its transmission a friction-drive. Although priced low, it couldn't compete with the *Ford Model T* and 1910 was the company's last year.

The *Kansas City,* built in Kansas City, Missouri in 1906, was offered as a 2-cylinder runabout and a 4-cylinder touring car. The next year the 2-cylinder was dropped, and the company added a 75-horsepower 4-cylinder car. In 1908, the car was called the *Kansas City Motor Buggy.* The company had financial difficulties in 1908 and in 1909 was restructured, to no avail, as the **Wonder Motor Car Company.** The next year the **Kansas City Vehicle Co**. moved

into the factory to produce their car, the *Gleason.*

1905 Kansas City Two Passenger Runabout

─────────────

In 1909, the **W.A. Salter Motor Company** was organized to build the *Salter,* a 4-cylinder,

30-horsepower vehicle. They moved into their factory at 1516 Oakland Street, Kansas City,

Missouri in September 1909.

The cars were equipped with F-Head, 4-cylinder, 40 horsepower engines. The roadsters

sold for $1,700, and the larger touring car for $50 more. Salter only produced a few cars each

year. However, the profits just weren't there and in 1915, Salter stopped making autos and

converted his factory into a machine shop.

─────────────

In 1908, the *Simplo,* the auto made by carriage-maker, the **Cook Motor Vehicle Co.,** was

advertised as "the biggest auto value in America." The St. Louis, Missouri car was a high-

wheeler, offered in several models, including runabout, roadster, and surrey. Solid tires were standard, but pneumatic tires were an option at $50 more. The vehicle had right-hand steering, a friction transmission, double-chain drive, and the 2-cylinder engine came in water- or air-cooled variations.

In 1909, the company gave up selling its own vehicle and became a car dealership, selling a variety of car makes.

The **All-Steel Car Co.** of Macon, Missouri, built a four-seater vehicle from 1915 to 1917 that was said to be like the *Scripps-Booth* automobile. A light vehicle, it used a 4-cylinder, 2-litre Sterling engine.

The *Ohio* was built, surprisingly enough, in Ohio. It was in Carthage, Ohio, to be specific, in the year 1909, that the **Jewell Carriage Company** reorganized as the **Ohio Motor Car Co.** and began producing a 4-cylinder vehicle available in roadster and touring models.

When 1911 rolled around, they decided to name each model. The *Mud Hen* was one. The company was in trouble when 1912 came and Ralph Northway, flush with cash after selling

his **Northway Engine Co.** to **General Motors,** bought the company and renamed it the **Crescent Motor Car Company.**

They never did get to manufacture the 1912 line, which would have had the following models: the *Euclid Torpedo,*

130

Grand Prix Bullet, Brighton Beach Speedster, Grandin Tonneau, and others.

———————————

In Xenia, Ohio, in 1914, Paul Hawkins started the **Hawkins Cyclecar Company** to build his cyclecar. The *Xenia* had a 2-cylinder, air-cooled engine, planetary transmission, belt-drive, and two, tandem seats. Selling for $385, the price was right, and the car was considered well-made. It won a few races, and one made the trek from Ohio to San Francisco without mishap, but its days were numbered. As the cyclecar "craze" petered out, the car was said to be unable to withstand the bumps and crevices of the roads of the day. They went out of business at the end of 1914.

———————

The *Cino* was built in Cincinnati, Ohio by **Haberer and Company**, from 1909 until 1913. It had a 4-cylinder engine. Starting with five body types, the marque did not prosper. It's last year it had only two, a 5-seater touring car, and a roadster.

Originating in 1904, the **Oscar-Lear Automobile Co.** of Springfield, Ohio used an air-cooled engine. It had a rotary blower which forced air into aluminum jackets that went around the engine's four-cylinders.

In 1905, the 24 horsepower auto sold for $2,300 and the company lasted until 1910.

Earl Sherbondy assembled his first gasoline engine in 1904 at the age of 16. He continued to study engines and by the time he graduated from University School in Cleveland in

1906, he had developed a two-stroke, single-cylinder, water-cooled engine. In 1908, the 20-year-old Sherbondy formed the **Simplex Manufacturing Company** in Cleveland to build engines and transmissions.

In 1909, the company brought out its automobile, a 4-cylinder, 30-horsepower, 7-passenger touring car, the *Derain,* and changed the company name to the **Derain Motor Company.** The cars sold for $4,000 each, which is possibly the reason the company was out-of-business by 1911.

The **Victor Automobile Company** manufactured the *Senator* in 1906, a 4-cylinder, air-cooled engine drove the car that was available as either a touring car or a roadster. The $2,000 car was manufactured from 1907 until 1910 at their factory at 4th and Portland Street in Ridgeville, Indiana. A smaller 14/16 twin was added in 1909 at $650, but it wasn't enough to save the company, which went under in 1910.

_____

The **Ohio Motor Vehicle Company** of Cleveland produced the *Ferris* auto from 1920 until 1922. The auto used a Continental 6-cylinder engine with an emphasis on excellent coachwork. Open and closed models were available, and disc wheels were standard equipment. By the time they closed shop in 1922, less than 1,000 autos had been sold.

_____

The **Lawrence Stamping Company** doesn't sound like an auto company, any more than *Odelot* sounds like an auto brand name. But in 1915, in Toledo, Ohio, it was all true. The name of the car, the *Odelot,* was Toledo spelled backwards.

The car was a 2-seat Raceabout with a 4-cylinder, 20 horsepower engine and sporty wire wheels. The company promised a touring model for January of 1916 but weren't around to keep their promise.

Another Ohio car company named MacDonald was the **MacDonald Steam Automotive** Corporation of Garfield, Ohio. In 1923 and 1924 they manufactured a vehicle named the *MacDonald Bobcat* that failed to prosper. A steam vehicle, the company was more successful selling the steam engines than the vehicle.

In 1918, the **Holmes Automobile Company** of Canton, Ohio produced the unusual looking (it was said to look like a "caterpillar head") *Holmes.* The auto had a 6-cylinder, air-cooled engine. The company produced about 500 cars per year, until 1923, when they declared bankruptcy due to the vice-president's embezzling and larceny.

Benjamin Gramm (not to be confused with Marvel Comics similarly named "The Thing") manufactured the *Logan* motorcar through his **Motor Storage and Manufacturing Co.** in Chillicothe, Ohio, in 1904. The *Logan,* a 2-cylinder, 10-horsepower vehicle, was said to be "perfectly balanced and vibrationless." The ride was so smooth that you could "easily write a letter in the car."

In 1906, a popular model was the 4-cylinder, *Blue Streak Semi-Racer Runabout* that sold for $1,750 and was so fast it could "feed its dust to anything on wheels." The slogan for the car was "The Car of Quality," which was also used by **Premier** of Indianapolis until Gramm sued them.

When the company went bankrupt in 1908, and Gramm moved to Bowling Green, Ohio, where he manufactured trucks until World War II as the **Gramm-Logan Motor Car Company.**

The Mahoning Motor Car Co. of Youngstown, Ohio, built the car named the *Mahoning* from 1903 to 1905. The primary vehicle had a 4-cylinder, 28-horsepower, air-cooled engine, and was chain-driven, with three-forward speeds.

Chapter Thirty-Five—The Orphans of the Atlantic States

The *Kline Kar* started in York, Pennsylvania, in 1910, and was enticed to move its operations to Richmond, Virginia in 1912, where it stayed until it closed in 1923. The company was incorporated as the **Kline Motor Car Corporation** in 1911.

The father of the company, James Kline, was known as a "master of the business." Their most notable car was the 650 Model Runabout. The *Kline Kar*s were expensive at $2,585, and were preferred by many Washington, D.C. politicians. In its short history, over 2,500 cars were built.

The *Biddle* was a well-regarded auto from R. Ralston Biddle, of the socially prominent Biddle family. Forming the **Biddle Motor Car Co.,** Biddle manufactured the small, 4-cylinder luxury car from 1915 to 1923 in Philadelphia. The cars were known for beautiful coachwork, and a wide variety of body variations. Few were built after 1921, and after 1923 the marque disappeared completely.

J. Walter Christie was an engineer and inventor. His *Christie* cars from 1904 until 1910 had front-wheel drive, which Christie was a strict proponent of. Most of these autos were built as racecars, purportedly to advertise the cars Christie would be making with his company, the **Direct-Action Motor Car Company** of 1906, which became the New York City-based **Walter Christie Automobile Company** in 1908.

The 1906 *Christie* was a large 50 horsepower touring car, with a 100-inch wheelbase, weighing in at 2,300 pounds.

Christie became the first American to race in the Grand Prix in 1907, with the largest car to ever take part in a Grand Prix. Later Christie vehicles included taxicabs, tractors, fire engines, and, for the military, tanks.

J. Walter Christie in his 1906 racecar.

Christie revolutionized suspensions with one he developed for tanks to use in World War I. Christie had many patents for all the inventions he had, and improvements on other's inventions. Even so, he often found himself low on cash, and famously said, "There is no shame in being poor, but it sure is damn inconvenient."

In 1914, the *Gadabout* of Newark, NJ and Detroit, MI was a 4-cylinder, 12 horsepower cyclecar with side-by-side seating (for two passengers). Magneto ignition, a distinctive "wickerware" body on a wood frame, and "splash lubrication" were a few features of the auto. (Wickerware was a design that looked plaited or like woven twigs or wicker.)

A factory was leased at Runyon and Badger Street in Newark and the company manufactured cars there for a year. In 1915, production moved to a factory in Detroit.

The company made an appearance at the 1916 Chicago Auto Show, but was purchased by Buffalo, New York's **Heseltine Motor Company** shortly thereafter and the *Heseltine,* named for company president Philip Heseltine, was built there. The *Heseltine* was a two-seater, 4-cylinder, 27 horsepower Runabout, and available in two wheelbase sizes; it was manufactured until 1917.

David Parry's two brothers didn't want to invest in autos, content with the profits from their carriage business. David Parry purchased controlling interest in the **Overland Motor Co.** in 1906. He was an inventor and sold out his **Overland** stock to John Willys, to finance his own auto plans. He got started in Indianapolis as the **Parry Manufacturing Company** and began manufacturing the *Parry,* a 4-cylinder runabout or touring car, in 1910.

In preparation, Parry leased seven buildings from the **Standard Wheel Company** and hired 348 employees in 1909. The name of the company was changed to the **Parry Motor Company**.

In 1911, the name of the car was changed to the *New Parry,* although the only real change was the higher price. By the end of the year, cash was running low and the 900 cars assembled and sold didn't help much. Although originally capitalized at a million dollars, only $150,000 had been paid in. Most of that had been spent on equipment and advertising.

As the company moved into receivership, creditors took over and reorganized as the **Motor Car Manufacturing Co**. In 1912, the new company name was added to the buildings, and the company continued to manufacture the *New Parry.* Later in 1912, a car named the *Pathfinder* shared the assembly line with the *New Parry.* By 1913, the *New Parry* was phased out.

The *Pathfinder* was a 4-cylinder vehicle, available in a variety of styles including *Touring, Phaeton, Armored Roadster,* and the *"Martha Washington Coach."* The car is widely regarded, and considered well-built, with a motto, "Known for Reliability." The car is also well-regarded because of the styling and colorful models that were available. They vowed to end the "drab days of color repression."

By 1914, a 6-cylinder was added. In 1915 the 4-cylinder models were discontinued in favor of the 6-cylinder models. In 1916, a 12-cylinder car was offered along with the 6-cylinder and the company name was revised to be the **Pathfinder Company.**

There were awards, including the Royal Automobile Club of England's Certificate of Performance. New York millionaire E. M. Pierce and his chauffer drove a *Pathfinder* auto 10,000 miles across the U.S. without any "distress" at all.

1915 Pathfinder

The post-World War I depression was one of the main reasons the company went under in 1917 (despite rumors it would merge with the **Empire Motor Car Company**). The factory was sold to an outfit that manufactured shoe polish.

Cunningham Auto Co. Sales and Service—along with Dort and International Motor Trucks

The *Cunningham*, like the *Franklin*, was considered a luxury car. Built in Rochester, New York from 1907 to 1936, it was one of the longest-lived car companies that met its demise in the Depression.

The 1922 Cunningham

James Cunningham Sons & Co., Inc. were one of the grand old firms that made carriages in Rochester, New York, before becoming a car company. At first the *Cunningham* auto was merely an assembled vehicle, acquiring parts from various specialty companies, such as 4- and 6-cylinder engines from Continental and Buffalo engine companies. But as the car progressed through the years, more and more of it was produced in-house.

The company continued to make carriages along with autos until 1915, when carriage production ceased and only the cars were manufactured. In 1915, a single *Cunningham* model was produced, with a V-8 engine and the usual fine styling.

Cunningham owners included William Randolph Hearst, Mary Pickford, Harold Lloyd, and Marshall Fields.

From 1935 to 1936, the company only made bodies for other manufacturers, and ambulances and hearses.

The **Delling Steam Motor Company** from West Collingwood, NJ, was founded by Eric Delling with his brothers as his partners. They sold the *Delling Steamer* from 1923 to 1927, although less than 100 of the 2-cylinder steam cars were made and sold.

The **Rainier Motor Company** originated in Flushing, Queens New York in 1905. John T. Rainier's car was a 4-cylinder, 35/40 horsepower vehicle available in touring and town car models.

The **Rainier Motor Company** had an arrangement with **Studebaker** to make a chassis. When **Studebaker** switched to **Garford**, Rainier moved the factory to Saginaw, Michigan, at 6th and Washington Street, in 1907. They continued to manufacture *Rainier*s until 1912, when the factory was purchased by **General Motors**.

The *Keystone Six* was built in Yonkers, New York, and Du Bois, Pennsylvania in 1909. Thirty of the cars were built in the **Howard Motor Works** facility in Yonkers while waiting for the Du Bois factory to reach completion.

The company, the **Munch-Allen Motor Car Company**, moved to the Du Bois factory. The car was a 6-cylinder, available in roadster, baby tonneau, or touring car styles, selling for $2,250, regardless of which style one chose.

In 1910, C.P. Munch brought the company back to Yonkers to no avail, 1910 was their last year.

The **Torbensen Motor Car Company,** founded by Viggo Torbensen in Bloomfield, New Jersey in 1902, advertised but produced few cars. Torbensen had started by making and selling auto specialties in Newark, New Jersey.

In 1901, he decided to specialize in producing gears, as **Torbensen Gear, Inc.** and moved to Bloomington, New Jersey. In 1902, he assembled a small runabout for his own use. He produced more cars over the years, the *Torbensen,* and sold them locally until about 1911.

Some of his creations were delivery and other commercial vehicles, usually with 6-cylinders, although a 3-cylinder delivery wagon was constructed. In 1912 he had stopped assembling the cars to concentrate on manufacturing automobile axles.

Swiss engineer Martin Fischer developed a slide-valve engine with sleeves that had both reciprocal and oscillating motion. The engine was called "the Magic" and was shown at the Berlin Automobile Show in 1911. Fischer began granting licenses and the first U.S. one was granted to **Aristos Industries** of 250 W. 54th St. in New York City, in 1914. They were producers of *Mondex* automotive products, including body polish, and shock "preventers."

They named their 6-cylinder touring car of 1914 and 1915 the *Mondex Magic.* They offered it in a 40-horsepower, $4,500 version, and a 60-horsepower, $6,500 version.

The venture wasn't profitable, and the next year the **Aristos Industries** went back to making their other *Mondex* products.

———————

J. William Jenkins of Rochester, New York, started off in the shoe manufacturing business. He manufactured the *Jenkins* from 1907 until 1912 as the **Jenkins Motor Car Company**.

The *Jenkins* was a 4-cylinder auto with a distinctive curved hood. Other than changing the size/wheelbase of the auto each year, the car didn't change much with each year. When Jenkins tired of the auto industry, he sold the company to his chief engineer, Fred Decker, who turned the business into a *Cole* dealership.

Jenkins auto, a 4-cylinder car that was built from 1907 until 1912, not changing much each year.

Klink Motor Car Co. of Dansville, New York, produced the *Klink,* with a 4-cylinder, 30 hp engine, from 1907 until 1909. For 1908 a 40 hp engine was offered. In 1909, a six-cylinder auto joined the fray. All was for nothing, as production ceased in September 1909 after only 20 vehicles were produced.

———————————

From 1900 until 1904, the **Conrad Motor Carriage Company** of Buffalo, New York, made steam cars with tiller-steering and single-chain drive, powered by 8 to 12 hp, 2-cylinder engines. They changed to 2-cylinder gasoline models in 1902, offering a runabout at $750, and a $1,250 touring car. The Founder and President Schuylar Fisher died in 1903, and the company died the year after in 1904.

Christian Weeber started building and selling bicycles in Albany, New York. In 1898, he began assembling his first auto, the *Weebermobile,* a single-cylinder, air-cooled vehicle with a sliding gear transmission and chain drive. He made a few more until 1905 when the **C.F. Weeber Manufacturing Works**, at 170-172 Central Avenue in Albany, started concentrating on building auto parts, auto repair, and as dealers for other autos, including **E-M-F, Ford**, and **Studebaker**. **Weeber** had many patents, including ones for a muffler, tires and a fuel inductor.

The **G.J.G. Motor Car Company** of White Plains, New York, manufactured cars from 1909 to 1911. Designed by G.J. Grossman, the car names included *Scout, Carryall, Comfort,* and *Pirate.* With a 4-cylinder engine, it was claimed these autos could attain speeds of 65 miles per hour. The hood had a distinctive "cupola" on the cars.

The **Palmer & Singer Manufacturing Co.** of Long Island City, New York, manufactured luxury 4- and 6-cylinder cars from 1907 until 1914. The 1913 models were called *Brightons.* When the company failed in 1914, it was succeeded by the **Singer Motor Car Co., Inc.**

From 1915 until 1920, the **Singer Motor Car Co., Inc.** manufactured autos with Herschel-Spillman 6-cylinder engines, distinctively-styled, and available in a wide choice of custom bodies. Wire wheels were standard, and prices went up as high as $9,000 for the vehicles. In 1920, the company's last year, a 12-cylinder engine was introduced.

The **Close Cycle Co.** of Olean, New York, became the **Close Cycle and Automobile Company** in 1902 when the proprietors, the Close Brothers, started making cars. It's not known exactly how many autos they made, but they quit in 1907 and became the **Olean Garage Co.** to concentrate on auto repair.

In Goshen, New York, Joseph Coates bought a racetrack called the "Historic Track" in 1884, and it was said to be the "birthplace -of sulky racing." A sulky was a two-wheel, lightweight vehicle for one person, to be driven by one horse. With the **Miller Cart Company**, Coates produced high-quality sulkies. In the sulky factory Coates also built a 4-cylinder experimental car in 1905. In 1908, he raised $150,000 financing to build a factory at 183 Greenwich Avenue in Goshen and started manufacturing the auto.

At the 1909 Automobile Show at New York's Crystal Palace, Coates unveiled the *Coates-Goshen,* a 4-cylinder, 32 horsepower vehicle with a 116-inch wheelbase. A smaller, 25 horsepower with a 112-inch wheelbase was also available, both available in Town Car, Runabout, and Baby Tonneau styles. The 1910 vehicles included 40- and 60-horsepower models.

When about 32 vehicles had been built, the town of Goshen suffered a terrible fire which burned down the factory, a church, an eight-story tenement, a lumberyard, and several private homes. In 1911, the company declared bankruptcy.

In 1912, Joseph Coates marketed a $450, 2-cylinder, three-wheeled commercial van named the *Coates Tricar* for two years and then, living into his 80s, he designed race tracks.

One of the *Coates-Goshen* buildings was repurposed into the Healey Brothers Chevrolet-Buick dealership building.

The 1909 Coates-Goshen

The *HAL* was named for the initials of its founder, H.A. Lozier. The New York auto firm introduced a 12-cylinder, Weidely engine in 1916. Two, four, and seven-seater cars were built, as well as limousines, priced at $4,500. The **Hal Motor Car Co.** went out-of-business in 1918.

In 1904, a consortium of upper New York businessmen formed the **Duquesne Motor Car Company** and then changed the name to the **Duquesne Construction Company.** The new company took the offer of Jamesville, New York, to start production of their car there, for a cash bonus of $5,000. They manufactured the *Duquesne,* a 4-cylinder, a 5-seat touring car with shaft drive and a round radiator. Other features of the *Duquesne* included headlights that turned with the steering wheel and an advanced starting system (without a crank)—for $2,000.

In 1906 the company went bankrupt and paid its stockholders back 10% of what they'd paid in. The factory equipment was sold to the factory mortgage holder. The total production was probably 6 cars.

———————————

The *Babcock* was manufactured from 1909 to 1913 by a Watertown, New York carriage-maker, the **Watertown Carriage Company.** The 2-cylinder, 18 horsepower vehicle was at first of high-wheeler design, but later models were a more conventional design. In 1913, founder George Babcock announced discontinuance of their auto line to concentrate on making truck bodies. By 1918, the company employed over 800 people, building ambulance bodies for World War I field action.

———————————

The **J.S. Leggett Manufacturing Co.** of Syracuse, New York manufactured the 1904 *Iroquois*, "a small car of advanced design." Lasting until 1908, the car had a sliding-gear transmission and shaft drive, progressive for the time. Originally 20 horsepower, later models offered 30 or 40 hp.

In Buffalo, New York, the **Kensington Automobile Co.** made gasoline, steam, and electric-powered cars from 1899 until 1904. The electric and steam cars were runabouts of similar body design. Both were two-seaters and had a 2-cylinder, 4 -horsepower engine with a single-chain drive. The gasoline-powered car debuted in 1902 and had a 2-cylinder, 11-horsepower engine.

In 1902, all three types of vehicle, steam, electric, and gas, were produced by the **Kensington Automobile Co.** The next year, 1903, the steam vehicle was discontinued. The next year, all production ceased.

From 1901 until 1903, the **Ward Leonard Electrical Co**. released both an electric vehicle and a gasoline powered one. The first was the *Century Tourist,* a 2-cylinder, 3 ½ horsepower electric vehicle, and one named the *Knickerbocker,* a gasoline-powered 2-cylinder vehicle. The first vehicle was retroactively renamed *Knickerbocker I,* probably because there was already an electric vehicle named *Century* being manufactured in Syracuse, New York. The **Ward Leonard Electrical Company** was in Bronxville, New York.

The company was led by Ward Leonard, who went on to have over 100 patents dealing with electricity. Besides developing electricity infrastructure in other areas, Leonard developed an auto lighting system. His company is still in the electrical business.

From 1917 until 1922, the **American Motors Co.** of New York City, built the *Amco,* a 4-cylinder car, available in either right-hand or left-hand drive because they were mostly intended for export to Europe. Described as "a mainly British car" and made of "the finest grade American components." Most of the cars were painted beige and cost $1,600. Not to be confused with the 1970's era **American Motors Company**.

Balzer Cyclecar

In the Bronx area of New York City, the *Balzer* auto was built by the **Balzer Motor Carriage Co.** from 1894 to 1900. The vehicle, like a cyclecar, used a 3-cylinder, rotary engine mounted vertically. Designed by Stephen Balzer, and the Balzer engine was used in the 1906 *Carey*. Of the several cars made, one survives in the Smithsonian Institute's collection in Washington, D.C.

The auto known as the *Pittsburgh 6* was actually manufactured in New Kensington, Pennsylvania in 1908 until 1910 when it moved to Pittsburgh. It received its financing when insurance agent H.M. Schmitt sold his agency, using the proceeds to finance the company he became president of, the **Fort Pitt Motor Manufacturing Co.** The company assembled 60 and 75 horsepower, 6-cylinder vehicles in roadster and touring car models.

In 1911, the company secured the plant of the former **Pittsburgh Steel Pulley Co.** and continued to assemble vehicles but went into bankruptcy after producing less than 20.

––––––––––––

Edgar Huselton was an auto dealer, the first in Butler, Pennsylvania. He sold *Maxwell, REOs,* and *Reliance* vehicles. In 1908, as **Huselton Automobile Co.,** he started building his first auto along the lines of luxury cars of the day, like *Pierce-Arrow* and *Packard*. He used the highest-grade parts of the day, such as Bosch parts.

His car dealership kept him busy, but he managed to release his first auto in 1911. Large, with a wheelbase of 123 inches, it had a 4-cylinder engine that could generate 40-horsepower. He manufactured autos off and on, roadsters and touring cars, until he quit in 1915 to return to his car dealership full-time. He had produced at least 13 autos and one ¾ ton truck from 1911 to 1914. His grandson has the last remaining *Huselton*.

Charles Middleby took over Charles Duryea's former **Duryea Power Company** factory in Reading, Pennsylvania for his **Middleby Automobile Company**. His auto was the *Middleby*, which was a 4-cylinder, 25 horsepower vehicle with a 108-inch wheelbase, available for $1,000 in 1909. By 1910 the car was available in six different styles.

Its first year it was offered in three models, *a Runabout, Surrey,* and *Touring Car*. The next year, 1910, *Single* and *Double Rumble Roads* and *Toy Tonneau* models were added. In 1911, the wheelbase was enlarged to 122 inches, and it had large, 36-inch wheels.

With a companion marque called the *Reading*, the business produced about 400 vehicles a year until its demise in 1913.

In 1919, Duncan MacDonald had patents for steam vehicles, and the **Gearless Motor Corporation** of Pittsburgh decided to build them. The cars had "two individual, double-acting, side-valve, 2-cylinder engines, which together can generate 65 horsepower."

In 1919, a 5-passenger touring car selling for $2,600 and a 2-passenger roadster at $2,650 were the company's offerings. The cars were available with either wood or wire wheels. "No claims, no knocks, just demonstration" was the slogan.

In 1920, Duncan MacDonald abruptly quit and moved to Gabriel, Ohio, ostensibly to manufacture cars under his own name. In the meantime, a few of the board members thought that stock malfeasance was more lucrative than auto manufacturing. Four of them, including Duncan MacDonald were indicted for conspiracy and using the mails to defraud. Over $1,650,000 worth of worthless stock was sold. In January of 1924, all four were found guilty.

In 1905, the **Chalfant Gasoline Motor Company** of Lenover, Pennsylvania, released its auto with a 2-cylinder, 22 horsepower, water-cooled engine. Production ceased in 1912.

The *Penn* was an auto manufactured by the **Penn Motor Car Company** of 7510 Thomas Blvd., Pittsburgh and New Castle, Pennsylvania, from 1911 to 1913. The *Penn 6* had a 4-cylinder engine, available in a 2-seater roadster or 5-seater touring model on a 105-inch wheelbase. In 1912, a 45-horsepower model with a 115-inch wheelbase became available.

Also in 1912, the company announced a move to a new factory being built in New Castle, PA. However, the factory owners changed their mind once the **Penn Motor Car Company** moved there, forcing the company to go into receivership in 1913.

The *Champion* was originally sold under the name *Direct Drive* in Pottstown, Pennsylvania, in 1917. The early models had the gearing mounted on the rear wheel rims, and later models used the more conventional transmission. Two similar models were available, the first was the *Tourist* with a Lycoming 4-cylinder engine and a *Packard*-like radiator shape. The other was the *Special,* with a 4-cylinder Herschel-Spillman engine and a radiator that resembled the *Rolls-Royce.*

In 1919, they released a 6-cylinder engine in the *Model C-6.* When the company was first founded, they were called the **Direct Drive Motor Co.** When they changed the name of the car to *Champion,* they changed the name of the company to **Champion Motors Co.** Prices ranged from $1,050 to $1,195.

In 1910, the **Louis J. Bergdoll Motor Company**, headed by a wealthy Philadelphia, Pennsylvania auto enthusiast, built some of the most well-regarded autos of their day. The other brothers, Grover and Erwin, were successful racecar drivers.

The 30- to 40-horsepower, 4-cylinder vehicles called a *Bergdoll* were built in a 100,000-square foot factory in downtown Philadelphia at 16th and Callowhill Streets. The factory, built of reinforced steel and concrete, was seven stories tall. Unfortunately, the company went into receivership in March 1913 and was auctioned off for $45,062.

From 1897 to 1900, the **Crouch Automobile Manufacturing & Transportation Co.** of New Brighton, Pennsylvania, produced an 8 hp, V-twin steam car. They didn't last much longer than the **Belden Motor Car Company**, who produced a 1907 catalog showing their upcoming 6-cylinder cars named the *Belden*. By 1909, they admitted they had just manufactured their first *Belden*. By 1911, this Pittsburg company was history.

The **Bell Motor Car Company** of York, Pennsylvania, manufactured a small car, the *Bell Model 16*. They used Continental engines at first, in 1916, and then Herschel-Spillman after 1919. Their cars had a 112-inch wheelbase and sold for $775. In 1916 they had the motto, "a four-cylinder car of many attractions" and used the slogan, "light and mighty."

Their peak year was 1919, when they sold over 500 autos. It was downhill from there as they tried to sign up more dealers. They were one of the first car companies to pursue African-Americans to start dealerships. By 1921 they were out-of-business.

G.E. Daniels was the President of **Oakland Motor Co.** when he moved to Reading, Pennsylvania to start his own company in 1915. The *Daniels* was a large, and expensive car, usually made to order. The autos were powered by side-valve V-8 Herschel-Spellman engines until 1919, when they started manufacturing their engines in-house. The 1919 *Submarine Speedster* and *Submarine Speedster* were especially well-renowned. The *Daniels* carried no identification except for the letter *D* on the hubcaps. Around 2,000 were manufactured in all.

In 1923, the company was purchased by **Levene Motor Company** from Philadelphia. They upped the price by $2,500 to $10,000 and assembled a few more cars until 1924.

The **Hanover Motor Company** of Hanover, Pennsylvania produced a cyclecar from 1921 to 1927. The car had a 2-cylinder, 15 horsepower air-cooled engine. Unlike most cyclecars made in the U.S, the car was intended mainly for export and marketed in Japan, where it was claimed that 800 were sold. With the choice of right or left-hand steering, it sold for $345.00.

In 1925, the old factory of the **Parentis Motor Corporation** in Buffalo was purchased for $225,000. Whether the company actually moved there is doubtful, and they were out-of-business by 1927.

The **York Motor Company** of Pennsylvania built the *Pullman* auto from 1905 to 1917, a large, expensive car with a 4-cylinder, 40-horsepower engine. In 1909 the company was reorganized and renamed the **Pullman Motor Company.** Billed as the "Palace Car of the Road," it won the Fairmont Park Road Race and was awarded three gold medals at the Russian Exposition. The marque lasted until 1917, after producing more than 20,000 cars.

_____

The **Standard Steel Car Co.** of Butler, Pennsylvania built the *Standard* with a 6-cylinder, 38 hp engine from 1912 until 1923. Available in touring and closed models, it sold for up to $3,600.

In 1916, an 8-cylinder model, smaller than the 6-cylinder, cost only $1,950 and had 29 hp. The next year the horsepower was increased to 34 hp. In 1921, prices were back up to $5,000. In need of restructuring by 1923, the company was renamed **Standard Auto Vehicle Co.** but was out of business by the end of the year.

————————

Another Pennsylvania car named *Standard* was produced in Philadelphia in 1910 by the **Standard Gas Electric Power Co.** The single model released had a 4-cylinder engine and a 3-speed, sliding-gear transmission and shaft drive, and was a 4-seater torpedo model. It featured one of the first auto electrical starters.

The **Kearns Motor Buggy Company** of Beavertown, Pennsylvania manufactured the *Eureka Buggy* in 1908 and 1909, a high-wheeler with a 2-cylinder, 12 hp engine and double friction drive. In 1910, the company concentrated on trucks until 1914, when they manufactured the *Lulu,* a cyclecar with a 4-cylinder engine. In 1915, they began making a light 18 hp car with a 4-cylinder engine, 3-speed gear box, and shaft drive. They also used the name **Kearns Motor Truck Company.**

The *Gurley* was a 1-cylinder, 2-seater buggy with bicycle wheels and tiller steering. It was built by Tom Gurley in his Meyersdale, Pennsylvania bicycle shop in 1900. Gurley was a jeweler and bookseller who sold bicycles (and assembled automobiles) on the side. Hoping to charge $600 per vehicle, he found he couldn't make a profit on it unless he charged at least $1,000. Gurley later went into partnership with his brother Oscar, establishing one of the first auto dealerships in Pennsylvania.

The 1913 Victor, built in Greenville, S.C.

Chapter Thirty-Six: Orphans of the West

In 1920, a car named the *Texas* was built by the **Texas Truck & Tractor Company** in Dallas, Texas. From 1920 until 1922, the Kansas City, Missouri-based **Wharton Motor Company** built their car the *Wharton* in Dallas.

In 1915, it was announced in the auto magazines of the day, such as *Horseless Age,* that the newly-formed **Texas Motor Car Company** from San Antonio was looking for investors and factory space to build their auto. They also announced that prospective stockholders could get a ride in the vehicle. Apparently, there weren't enough takers since the company was never heard from again.

In 1920, the **Texas Motor Car Association** sold over 200 units of their car named the *Texan.* They were assembled in Fort Worth. The company went into receivership in 1922 with 100 unfinished cars in stock, and $500,000 worth of parts and $200,000 worth of tires.

In 1995, a new **DeLorean Motor Company** was created in honor of the original company started by John DeLorean in the 1980s to produce his iconic, gull-winged door vehicles. The current **DeLorean Motor Company** is in Humble, Texas, near Houston. They acquired the remaining parts inventory of the *DeLorean* to support current *DeLorean* owners. They have also purchased the rights to the stylized DMC logo of the original company.

Auto dealer Chester and druggist Ellsworth Jones, brothers from Beatrice, Nebraska, started their **Jonz Automobile Company** with the 2-cylinder auto Chester had manufactured. They readied it for the 1909 Chicago Auto Show but unfortunately, they ran out of money before they could complete it. They received financial help from another car dealer, as well as a lawnmower company, and the Beatrice newspaper even ran an appeal for people to contribute to the Jones auto project, all to no avail. Finally, they found an investor, Berton B. Bails, from Louisville, Kentucky, with whom they reorganized as the **American Automobile Manufacturing Co.** and moved all their equipment from Nebraska to a new factory in New Albany, Indiana, just across the Ohio River from Louisville, KY.

Perhaps the greatest achievement of the company was the care they put into their prospectus, in their quest for new stockholders. Glossy, with pictures of just about everyone in the company in it, they were able to entice over 8,000 people to invest.

The car was advertised as having a "vapor-cooled" engine, and no valves, no push-rods, no gears, no rollers, no rocker arms, no pumps, no radiator, and no water! It was called the *"Tranquil Jonz."*

One investor went to the factory to view a test drive and said the car went about a hundred yards and came to a sudden, but tranquil, stop. Continental engines were brought in to substitute for the substandard ones previously used, but to no avail. In March 1912 the company went bankrupt. The remains were bought by Fred Kahler in January 1913 and the factory was used to make his car the *Pilgrim*.

———————

The **Oklahoma Auto Manufacturing Company** made oil-field trucks in North Muskogee in 1915 until the Great Depression, when they went under.

———————

In 1906, 17-year old Harry House got some press attention from the Cheyenne, Wyoming newspapers. He was working on a car with a motorcycle engine and various bicycle parts. Nothing more was heard after 1906.

———————

The name *Golden State* was a popular one for California car makers to use. Unfortunately, none of these companies were very successful.

The first vehicle named *Golden Gate* was nothing more than a large tricycle with a gasoline motor on it. Built in San Francisco in 1893, it was an early attempt at constructing a viable gasoline engine-driven vehicle.

The second *Golden Gate* auto was built a little south—in San Jose from 1902 until 1903. It was a 2-cylinder, 8-horsepower, friction transmission vehicle. It was a viable vehicle but wasn't produced in mass quantities as planned.

The third unsuccessful vehicle called *Golden Gate* was built many more miles down the coast, in Los Angeles. Former L.A. policeman Ross Philips had put together a 2-passenger runabout, powered by an air-cooled engine. In 1904 through 1906, the **Golden State Motor Car Company** struggled to produce the vehicle and get their company get off the ground. They were never profitable building autos but found success when they switched to making kitchen stoves.

In 1917, a group of Oklahoma businessmen formed a company called **the Midland Motor Car and Truck Company**. Their goal was to make a car called the *Oklahoma Six* and a truck called the *Ozark*. With a U.S. Army contract, the company began making trucks for the government. After World War I, the company had to be reorganized because two of their stock officers were convicted of embezzlement. Upon reorganization, the company was purchased by the Texas-based **Wichita Motors Company**. They produced the *Wichita* truck there until 1922.

From 1904 until 1906, the *Chadwick* was manufactured first at Chester, Pennsylvania, as the **L.S. Chadwick Co.,** then in Philadelphia in 1907 and 1908 as the **Fairmont Engineering Co.,** and then, finally settling in Pottstown, Pennsylvania from 1907 until 1916, as the **Chadwick Engineering Works**.

The 1904 *Chadwick* had a 4-cylinder engine with double-chain drive. For 1905 and 1906, the car was upgraded to a 4-speed progressive transmission. The 1907 *Great Chadwick Six* introduced the 6-cylinder engine and had copper water-jackets and overhead valves. This car was considered "high-performance" and was used for car-racing. The 1908 version entered the Vanderbilt Cup and the Savannah Grand Prize races. These cars used "supercharging" to increase the car's power. This was not included on the regular *Chadwicks.* In 1911, the wheelbase increased to 133 inches. The marque lasted until 1916.

From 1920 to 1921, the *Reese Midget AeroCar* was manufactured in Huron, South Dakota by the **Sheldon F Reese Company, Inc.** The car was small and only weighed about 150 pounds. The 2-seater had one seat behind the other—the total wheelbase was only 60 inches. It was powered by a 2-cylinder, 2-stroke, 6 horsepower engine, and cooled by an air-screw propeller mounted on the rear of the car. Its price was a meagre $160.00.

The Reese Midget AeroCar

The car was "valveless, springless, beltless, and gearless." The top speed was 40 mph and it could go 61 miles with its gallon and a half gas tank full.

For winter states, a pair of runners was available to turn it into an ice-boat. As an iceboat, it could go 60 mph.

In 1911, until 1914, the **Cleburne Auto Car Manufacturing Company** built the *Luck Utility* truck and the *Cleburne* auto in Cleburne, Texas. Founded by Reverend Henry Luck, it was designed primarily to be a car that could navigate the abysmal roads of the day. With 4-cylinders and 30 horsepower, the car received high grades at first, but production was a problem. By 1914, they had only produced about 20 vehicles and ran out of luck, despite a name change in 1913 to the *"Luck Truck."*

Chapter Thirty-Eight—The Canadian Car Companies

There were many Canadian car companies, even though all the Canadian provinces haven't always been auto-friendly. On Prince Edward's Island, autos were banned from 1908 until 1913. From 1913 to 1918, autos were only allowed on the island on Mondays, Wednesdays, and Thursday. Finally, after 1918, the ban was dropped entirely.

In 1899, brothers Milton and Dr. Nelson Good formed the **LeRoy Manufacturing Company** in Berlin (now Kitchener), Ontario. They manufactured about 20 vehicles, for $650 apiece, until 1904, when evolving technology caught up with them and they went out of business. Their car was a close copy of the *Oldsmobile Curved Dash* auto, with a gasoline-powered engine and 2-cylinder engine. The same pedal was used to put the vehicle into reverse as was used for the brake, causing confusion among consumers.

The LeRoy

The Toronto-based **Russell Motor Car Company** built high-end, solid, and well-engineered autos, starting in 1905. Made from mostly Canadian materials, the cars were expensive, but popular. The company stopped making autos and started making defense products in 1914, for World War I, and was sold to John N. Willys of **Willys-Overland Motor Company** a few years later.

The Russell Motor Car

Billy Durant, the founder of **General Motors,** had a partner in his earlier business, the **Durant-Dort Carriage Company**, J. Dallas Dort. Dort started his own auto company, the **Dort Motor Car Company** after the success of Billy Durant with **GM.** Robert Gray met with Dallas Dort and secured the Canadian rights to produce Dort's car, the *Dort.*

Named the *Gray-Dort*, the car manufactured two models the first year, 1915, in Chatham, Ontario, and were called the **Chatham Motor Company**. They were the *Model 4 Roadster* and a *Model 5 Touring Car* and were known for their reliability, including being able to start in all kinds of weather. By 1919, there was a sedan model (for $1,245) added to the roadster and touring models.

In the early 1920s, when Dallas Dort first left the company, and then abruptly died, the Canadian version of the company struggled and then went out of business.

In 1931, the **Durant Motors** conglomerate was faltering and soon to go out of business. A group of Toronto businessmen bought out the Canadian division to establish **Dominion Motors** and manufacture a 6-cylinder auto called the *Frontenac*. When it didn't sell well the first year, they tried a re-design. But coming out in the beginning of the Great Depression, it didn't fare well, and the company closed in 1933.

In 1906, the first **Chatham Motor Company** was organized by a group of Chatham, Ontario investors to build the *Chatham,* a 5-passenger touring car with a 20 horsepower, 4-

cylinder, water-cooled engine. Since the driving etiquette had not yet been established yet for North America, the steering wheel was on the right-hand side of the car. In 1907, a Detroit creditor sued the company into liquidation. The remnants were purchased by a Chatham dentist, who resumed production of the *Chatham* in 1908.

Although the new *Chatham 30* had a new 30 horsepower engine, it only sold about 30 to 35 cars—mostly to the well-to-do citizens of Chatham. The company was sold to the **Anhut Motor Company** of Detroit in 1910. They used the old company's Chatham factory on Adelaide Street, about halfway between McGregor Creek and King Street, to once more produce a car branded the *Chatham.* The factory had been first used by **Hyslop and Ronald Company** to build fire engines.

In 1974, Malcolm Bricklin manufactured his "high-performance safety car" the *Bricklin S-V 1,* in Saint John, New Brunswick. Hiring the man who designed the Batmobile, Herb Grasse, to design it, the *Bricklin* had severe engineering flaws and went out of business in 1976, selling less than 3,000 vehicles. He later had more luck importing *Subarus* and *Yugos* from Europe and selling them in North America.

The Tudhope family was known for manufacturing lumbering, mining, and agricultural equipment. One of their divisions was the **Tudhope Carriage Company.** In 1908, the Tudhopes arranged with McIntyre of Auburn, Indiana, to acquire mechanical parts, and the **Tudhope-McIntyre Company** was formed to manufacture autos. The auto division was headed by William Tudhope.

The firm made the conversion to autos by manufacturing a 2-cylinder, air-cooled engine on a high-wheeler body, with a 2-speed transmission. It was essentially a motorized carriage. The vehicle, called the *Tudhope-McIntyre,* sold well throughout Canada for $550, until 1909 when its factory had a major fire. The factory was rebuilt, but William Tudhope felt he needed to update the "buggy" look of the high-wheeler auto. He arranged with the **Metzger Motor Company** of Detroit to manufacture a version of their up-to-date *Everitt 30* automobile. (These were the same Everitt and Metzger of **E-M-F** fame). The 1911 models reflected this partnership with the *Everitt-Tudhope* models, manufactured by the re-structured **Tudhope Motor Company.**

The **Metzger Motor Company** was taken over and William Tudhope decided to go it alone in 1913. Although it added electric lights and starters, the car was not doing well sales-wise, probably due to its expensive $1,600 price tag. The company was taken over by Walkerville, Ontario *Studebaker* plant manager Frank Fisher and the car was briefly renamed the *Fisher*. This only lasted until World War I commenced, when the company was liquidated, and the remaining parts were used to build ambulances.

In 1917, the *Moose Jaw Standard* was manufactured in Moose Jaw, Saskatchewan. They had Continental engines and were considered luxury cars. The concern was started by five residents who arranged to have the parts imported from the United States. Once each person had assembled themselves the luxury car, no one else was interested in one and the investors sold off the rest of the parts.

The *Comet* was manufactured in Montreal, Quebec. Manufactured by the **Comet Motor Company,** King George V rode in one on his 1908 visit to Quebec. It was developed by bicycle racer Oscar Robertson and manufactured from 1907 until 1909, when about 50 cars were built. Since the cars were hand-built, it was hard to turn a profit and the company didn't have the finances needed to fund a larger operation.

Part Seven:

Automotive

Topics

Chapter Thirty-Eight—The History of 4-Wheel Drive

Four-wheel drive, also known as 4 x 4, or four by four, and often written as 4WD, is a type of vehicle, able to provide power and torque to all wheel ends of a two-axle vehicle simultaneously. Although there are exceptions, the 4 x 4 term is often used interchangeably with the all-wheel drive, or AWD description.

A steam engine, with the first four-wheel-drive system for a steam-powered traction engine, was developed in England in 1893, by Bramah Joseph Diplock. Globally, the credit for developing the concept of 4WD *vehicles* is given to Ferdinand Porsche, who later designed the *Porsche* and *Volkswagen*. In 1899, he designed an *electric* vehicle with 4WD for Austrian truck maker Jacob Lohner; the car was called the *Lohner-Porsche*. This car debuted at the 1900 World Exposition in Paris, France.

In 1903, the first four-wheel drive car driven by an internal combustion engine is said by many to be the Dutch vehicle, the *Spyker*. The inventors were brothers Jacobus and Hendrik-Jan Spijker of Amsterdam in the Netherlands. *Mercedes* (**Daimler-Benz**) also started building 4WD vehicles in 1903, and their cars had all-wheel steering as well.

In the United States, one of the first 4WD proponents was Robert Twyford of Pennsylvania. His company was the **Twyford Motor Car Company,** and in Pittsburgh and Brookville, Pennsylvania, they built the *Twyford* from 1899 until 1907. Besides having 4WD, it had the first crude power-steering, using bevel gears.

Twyford moved the factory from Pittsburgh to a 10-acre site offered by Brookville in 1904. The company moved into the 2-story, brick factory and continued to produce vehicles there until 1907. After the company folded, Robert Twyford continued to profit from his 4WD

patents.

1904 Twyford Stanhope, most likely the first American 4WD vehicle. It sold for $1,200.

In Mankato, Minnesota, candy-maker Ernest Rosenberger put together a 4-wheel-drive

vehicle in 1909. He felt 4-wheel drive would help navigate the hilly terrain. Forming the **Four-**

Wheel Traction Co. to manufacture the car, the town held a ceremony opening the new factory

in 1907. There were three workers in all, and they had assembled five cars by 1908.

By 1912, over 30 cars and trucks had been made. The factory continued building *Katos,*

(they were also called *Four Tractions* and *Mankatos*), until 1913, when the **Four-Wheel Traction**

Co. sold out to the **Nevada Manufacturing Co.** of Nevada, Iowa.

1910 Kato 4WD truck

Another major American venture into 4WD came in 1908, when inventor Otto Zachow designed his shaft-drive steam-powered *Z&B*, a four-wheel-drive vehicle, in Clintonville, Wisconsin. Zachow and his brother-in-law William Besserdich formed the **Four-Wheel Drive Automobile Company** (also known as the **Badger Four Wheel Drive Automobile Company**) in Clintonville to produce the vehicle. For a time, they were the world's largest producer of 4-wheel drive vehicles. Their 1908 car, the *Battleship,* was considered the first successful four-wheel drive (4x4) car manufactured for the public. Their patent, which was the basis for their 4WD vehicle, was for a double-Y universal joint, encased in a drop-forged ball- and-socket.

The company switched to making trucks exclusively and opened a division in Canada in 1918 and in Great Britain in 1921. During World War II the company manufactured over 15,000

4WD trucks for the American and British military. The company became **FWD Corporation** in 1958. They have since been incorporated into **ELB Capital Management.**

Frank H. Morse of Milwaukee, Wisconsin, built a 4-wheel drive steam auto in 1902, which he sold for $550. The next year he joined the **Four-Wheel Drive Wagon Company** in Milwaukee, working on manufacturing commercial vehicles.

In Lanark and Rockford, Illinois, a 1901 steam 4WD car was produced by the **Cotta Automobile Company.** Charles Cotta had submitted and received his patent in 1900. He sold his patent to the **Four-Wheel Drive Wagon Company** of Milwaukee, Wisconsin in 1903. They continued producing 4WD autos until 1906. Thereafter, Charles Cotta started a successful company (the **Cotta Gear Co.**) that built transmissions.

In 1909, two Bakersfield, California gentlemen, Morton Homer Magie and Charles Nelson Winters, patented a shaft-drive four-wheel-drive vehicle. They had allegedly been working on it since 1904. They formed the **Road Runner Auto Company** to build the car, but any actual manufacturing is doubtful.

Another Californian, Charles Van Winkle of Farmington, built a vehicle in 1904 that was a chain-driven, four-wheel-drive vehicle. He built at least one prototype and then sold the design to the **Stockton Four Wheel Drive Motor Co.**, which failed to manufacture any autos.

One of the most notable 4WD vehicles was released by the **Jeffrey/Nash Corporation** in 1913. The vehicle was called the *Quad* and had four-wheel drive, four-wheel brakes, AND four-wheel steering. Between 1913 and 1919, over 11,000 of the vehicles were manufactured. By 1928, total production topped 41,000 trucks.

The Jeffrey Nash Quad truck with 4WD.

Other notable 4WD vehicles and facts:

-In 1914, Jesse Livingood from Graysville, Pennsylvania, created a kit that would convert a *Ford Model T* into a 4WD vehicle.

-Douglas and Maurice Steiger developed the first successful 4WD tractor, the *Steiger Tractor*, in 1958.

-From 1936 to 1944, the Japanese company **Tokyu Kurogane Kogyo** manufactured 4WD military vehicles for the Japanese military. The vehicles had a transfer case that engaged the front wheels.

-In 1938 the Russian military built the *GAZ-61,* a 4WD military vehicle.

-During World War II, the **American Bantam Co.** started production on the 4WD *Jeep.* Further *Jeeps* were produced by **Ford** and **Willys**. The *Jeep* was manufactured by **Willys,** on to **Kaiser,** then **Nash Motors Co.,** and then **American Motors Company**, and is now a **Chrysler** vehicle.

1944 Willys Jeep

-After their success with the *Jeep,* the **Willys Co**. manufactured the *CJ 2-A* in 1945, a 4WD vehicle.

-The **Willys Co**. successor, **Kaiser-Jeep**, released the *Jeep Wagoneer* in 1963, a very successful 4WD vehicle with automatic transmission.

-Trucks and off-road vehicles are now often 4WD exclusively. In 1946 Dodge released the *Power Wagon* and in 1948 Ford released the *F-Series* pickup truck.

-In 1980, Chrysler releases the **Eagle** car division, to manufacture a 4 WD passenger car.

1946 Dodge Power Wagon

Portions first printed in the Thumb Print News.

Chapter Thirty-Nine—Everybody and His Brother Had an Auto Company

The most common bond among car company creators was one of brotherhood. That is, there are *many* car companies that were started by a pair of brothers.

Probably the first officially-recognized American car was the one built by the Duryea Brothers, Charles and Frank, who manufactured and sold a gasoline-powered car in 1893. Some of the more famous brother companies had "brothers" in their name, including the **Graham Brothers, Apperson Brothers,** and the **Dodge Brothers**.

Brothers Powel and Lewis Crosley produced a car in 1939. Louis, Arthur, and Gaston Chevrolet formed **Frontenac Motor Company** together and they were all race car drivers. The Havers brothers, Fred and Ernest, made the Haver in Port Huron. The Kissel brothers made luxury cars in Wisconsin.

There was James Ward Packer, who bought a *Winton* and had to fix it many times on his way home to Warren, Ohio. When he saw Henry Winton later and tried to tell him how to improve the car, Winton told him, "If you think you know so much, why don't you start your own auto company?" Which James Packard did, enlisting the help of his brothers Weiss and Hatcher.

The **Studebaker Brothers** had the largest buggy business in the U.S. when brother Johnny convinced brothers Henry and Clem to branch off into autos. The **Grinnell Brothers** produced an electric car until they decided there was more money in pianos.

The Briscoe brothers, Benjamin and Frank, were instrumental in the history of many early companies. These include *Buick* and *Maxwell*, which they invested in. With Ben at the helm, the Briscoes also formed the **United States Motor Company**, and a car called the *Briscoe*, produced in Jackson.

The Welch brothers, Allie and Fred, had the successful **Welch Motor Company** and the Apperson brothers built some of the first autos. Robert and Louis Hupp started the **Hupp Motor**

Company and later the **Hupp-Yeats Electric Car Co**. **The Imperial Motor Company** was founded by brothers T.A. and George Campbell of Jackson, Michigan.

Charles and Frank Duryea in their auto, and the Apperson Brothers in their 1894 Haynes-Apperson.

Other brother teams, most of which named their auto companies for themselves, included brothers with the last name of **Bartholomew, Bergdoll, Blood, Burns, Cameron, Christopher, Close, Colburn, Dolling, Duesenberg, Eckhart, Fisher, Good, Grout, Gurley, Huffman, Jones, Lane, Lyons, Matheson, McCormack, Moon, Owen, Packard, Pope, Roebling, Sommer, Stoddard, Woods,** and **Zimmerman.**

The Chevrolet Brothers, left to right, Louis, Arthur, and Gaston. All three brothers were racecar drivers.

Robert, Joseph, and Ray Graham, the Graham Brothers. They started off making glass jars and then trucks before buying Paige Motor Company.

John and Horace, the Dodge Brothers

The four Kissel Brothers in front of the family hardware store in the 1890's.

The Studebaker Brothers moved from carriages to automobiles.

Edgar and Elmer Apperson

Chapter Forty—The Previous Professions of the Automakers

Probably the most obvious profession that automakers came from is carriage-building. As cars became more popular, carriage makers could see that horse-driven buggies and wagons were on their way out. The largest carriage-makers in the U.S. were the **Studebaker Brothers** of South Bend, Indiana, The **Durant-Dort Carriage Company** helped turn Flint into the "Vehicle City" until both partners, after making carriages for over 20 years, both branched off into the auto field. Billy Durant started **General Motors**, while J.D. Dort started the **Dort Motor Car Company.** Carriage companies also had the advantage of not having to look for a factory to build their autos, since the carriage company would often be spacious enough to allow auto manufacturing, at least in the beginning stages for a company.

William C. Anderson had a carriage company in Port Huron, which he sold and moved to Detroit to start the most successful electric car company, the **Detroit Electric Car Company**, which lasted until 1940. **Rauch & Lang Company** bought out the **Buffalo Electric Carriage Company** and were successful for several years. These and many other carriage companies went into the auto business.

Other carriage-makers making the transition to autos included carriage and buggy companies **Tudhope, Brewster, Owen Brothers, Anderson, Gabriel, Imperial, Conrad, Cortland, Babcock, Drummond, Everitt, Deal Burns, Spaulding, Zimmerman, Eckhart, Aurora, Pratt, Andrews, Velie, Rauch & Lang, McFarlan** and more. They all started as carriage makers and switched to auto manufacturing. A few of them, like **Cunningham**, continued making carriages while simultaneously building autos.

The second most prevalent profession of automakers was bicycle-related, either manufacturing, selling in a store, or both. During the 1890s, bicycles were as popular as autos were in the 1910s and 1920s, and planes in the 1930s. Some of the classic car pioneers that started out making or selling bicycles were the **Duryea Brothers**, the **Dodge Brothers**, **Elmore Manufacturing Company,** Alexander Pope and the **Pope Manufacturing Company,** Elwood Haynes and the **Haynes Automobile Company**, the **Apperson Brothers, Copeland, Metzger, Keating, Columbia, Hanson, Rider-Lewis, Waltham, Waverly, Welch, Pierce Arrow, Weeber, Garford, Kirk, Shell, Peerless, Lozier, Chandler, Comet,** and **Duesenberg.**

Other professions that had a lot of people join the auto business were makers of farm implements or appliance manufacturers. and machine shops that made a type of consumer product or trinket.

Many of the early autos, especially those of the high-wheeler variety, were based on carriage designs.

Two sewing machine companies—**White Sewing Machine Co**. and **Singer**, manufactured cars. **Pierce Arrow** started off making bird cages, ice boxes, and bicycles before switching to cars. The **Matheson**s had a mail-order furniture business.

Chapter Forty-One—The Most Popular Auto Company Names

Many cars and car companies were named for their owners or one-time owners, such as Ford, Chevrolet, Buick, and Chrysler. But many other names were favored for auto companies, most of which never produced any cars, or more than a prototype.

One of the most common names was "Motor," as in **Motor Car Company, Motor Vehicle Co., Motor Touring Car Company, Motor Vehicle Co., Motor Vehicle Car Appliance Co.**, and about 50 more! There were two vehicles called *"Motor Chair,"* one from Chicago and the other from San Diego.

The real, easy, classy, comfortable, luxurious way to see and thoroughly enjoy the Exposition is in an Osborn Electriquette, which supplants the antiquated push-chair and jinrikisha. The only passenger conveyance permitted on the grounds.

The simplicity of operation renders experience unnecessary. A child can drive it. It's great fun.

Stations at each entrance gate, all prominent points, and "Neptune's Wonderland" on the "Isthmus."

EXPOSITION MOTOR CHAIR COMPANY,
San Diego Exposition

Another favorite name was "Eagle" and there were more than 30 car companies with Eagle in the name. "Page" was part of the **Page Motor Vehicle Co.** of Providence, Rhode Island, the model name *Page-Toledo* in the **Toledo Motor Company,** the *Page* of J.J. Page of Owosso, Michigan, and the *Page* of Adrian, Michigan, a prototype of the **Church Manufacturing Co.,** whose primary business was the manufacture of wire fences.

The **Pungs-Finch Motor Company** started in 1904 as the **Pungs-Finch Auto and Gas Engine Company** in Detroit. The first autos were assembled in a boat-building facility. They featured sliding-gear transmissions and shaft drives, revolutionary for the time.

W.A. Pungs was the money-man, and his son-in-law E.B. Finch was the designer/engineer. Arguments between the two eventually broke up the company, but they managed to stay together until 1910 and manufacture several hundred *Pungs-Finches.* The *Finch Limited* was the most desired, with 50-horsepower and guaranteed to top 55 miles per hour. E.B. Finch later became a *Chalmers* dealer.

Henry Ford's **Ford Motor Company** was successful and the number one producing automaker in the U.S. even before he released the famed Tin Lizzie, the *Model T*. After the heyday of the *Model T*, **Ford Motor Company** put out many famed and desired autos.

1910 Ford Model T Touring Car

Starting with the 1903 *Model A*, Ford sold enough cars to move into larger factories, first in the Milwaukee Junction, then in Highland Park, and then in the world's largest plant at River Rouge.

1903 Model A Roadster

1936 Ford

1935 Ford "Business Model"

Until **Ford Motor Company** put out the Model T and priced it low enough to interest the middle and lower classes, most autos were pointedly for the rich only, with features and a price that was aimed at the wealthier members of the U.S. Trying to compete in a very crowded market—the luxury car market—caused many of the luxury car companies to fail.

Something else the Ford Model T did, in 1908, was set the precedent for steering wheels to be on the left-hand side of the car, instead of the right. (This innovation was said to be suggested by Clara Ford, who didn't like to get out in the mud in the gutter on the right.) At first, in 1898 when **Jeffery** of Wisconsin replaced the tiller with a steering wheel, most of the auto companies' steering wheels were put on the right of the dashboard.

Since the 1792 law that stated that traffic, which at that time included horses, wagons, and carriages, should drive on the right side of the road, that is how American traffic has moved. Therefore, when cars were first being manufactured in the United States with a steering wheel, it was positioned on the right-hand side of the vehicle until Henry Ford released the *Model T* in 1908, with a steering wheel on the left side. Ford said this was to "make it easier for people entering on the passenger side to avoid oncoming traffic."

By 1910, most cars had adapted to the steering wheel on the left, except *Pierce-Arrow*, which didn't change until 1920. Stutz still had a couple of cars available with right-hand steering in 1921 and 1922.

———————

Chapter Forty-Two—More Auto Company Firsts

The advancements in auto technology were accomplished by many different auto companies. These are some of them (with some in dispute):

-First steering wheel in a mass-produced car was by the **Jeffrey Co.** *Rambler* in 1904—by 1914 the tiller was gone from all cars. This is also credited to the 1901 *Packard*.

-First self-starter-1912 *Cadillac*-also 1926, *Cadillac* also had the first heat in a car.

-In 1902 *Oldsmobile* had the first speedometer.

-First automatic transmission-1939 *Oldsmobile*-also first in-car GPS in 1995.

-First mass-produced front-wheel drive auto was the *Cord* of 1929.

-First seatbelt in a mass-produced car was in a *Nash* in 1948.

-First adjustable seats were in the 1921 **Hudson.**

-First pushbutton door lock was in the *Franklin* in 1917.

-First mechanical windshield wiper was in the **Willys-Knight** in 1916.

-First all-mechanical automatic transmission was by **REO** in 1933.

-First 4-speed gearbox transmission was by **Locomobile** in 1904.

-First rearview mirror was in the 1910 *Marmon Wasp.*

-First auto air-conditioning in 1939 was in the *Packard One-Eighty.*

-*Packard* also had the first power windows in 1941 and power locks in 1954.

-*The 1940 Packard 180 had the first auto air-conditioning.*

-First automatic transmission-1939 *Oldsmobile*-also first in-car GPS in 1995 and in 1902, Olds had the first speedometer.

--First mass-produced front-wheel drive auto was *Cord* in 1929 (Also the first vehicle with hidden headlights).

-The first seatbelt in a mass-produced car was in a *Nash* in 1948.

-The first car radio was in the *Chevrolet* in 1922, although some sources say the 1923 *Springfield.*

-The first hydraulic (rather than cable-actuated) four-wheel system came in 1921 on a *Duesenberg*

-The vacuum-boosted or "power" brakes first appeared on the 1928 *Pierce-Arrow.*

-The first mass-produced disc brake was in the 1949 *Crosley.*

--- The first car fitted with an alternator, rather than a direct current dynamo, was the 1960 *Plymouth Valiant.*

-The first car fitted with a replaceable cartridge oil filter was the 1924 *Cadillac.*

- The first car to use a standardized production key-start system was the 1949 *Chrysler.*

-First 4-speed gearbox transmission was by the **Locomobile Motor Car Company** in 1904.

-First back-up lights and molybdenum steel on a car was for the 1922 *Will Sainte Claire.*

-First adjustable driver's seat was in the 1914 *Maxwell.*

-First windshield defroster was in the 1928 *Studebaker* and the first windshield washer was in the 1937 *Studebaker*.

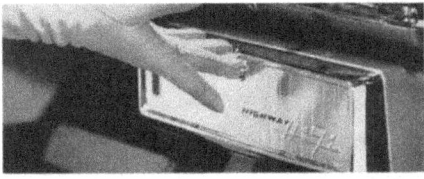

HIGHWAY HI-FI PHONOGRAPH
... provides the music you want wherever you go

NOW... another exclusive from Chrysler Corporation!

It's another Chrysler Corporation first! and can be operated without taking Highway Hi-Fi is just one of many
Highway Hi-Fi gives you the finest tone your eyes off the road. accessory new and exclusive features
reproduction... even on rough roads. A The 16-2/3 rpm records play from 45 that are available on all the cars of
special counter-balanced pick-up arm minutes to an hour on each side and the company. Look it. There's new
and shock-proof case mount unexposed are available in a wide variety of music magical Pushbutton PowerFlite
display. Conveniently located under the old and dramatic classics, popular new Lubriguard door latches new
instrument panel the Highway Hi-Fi favorites, music for children. famous auto hydraulic brakes all new
record player slides in and out easily Broadway musicals a store-com- FLIGHT-SWEEP styling. See all of
plete portable record library of effects. these new features at your dealer's now.

CHRYSLER CORPORATION ➤ THE *FORWARD* LOOK
PLYMOUTH • DODGE • DE SOTO • CHRYSLER • IMPERIAL

-The 1903 *Tincher* had the first power brakes.

-*Stutz* had the first safety glass.

-The first car fitted with a replaceable cartridge oil filter was the 1924 *Cadillac*.

-First gasoline auto with 4-wheel drive was the 1904 *Twyford Stanhope*.

-First steam auto with 4-wheel drive was built in 1902 by Frank Morse but not marketed.

--The first pushbutton door lock was in the *Franklin* in 1917

- The first car to use a standardized production key-start system was the 1949 *Chrysler*.

--The first auto to place the horn button in the center of the steering wheel was the 1915 *Scripps-Booth Model C*. This car also had the first *electric* door latches.

-The 1956 *Chrysler* had the "Highway Hi-Fi" record player.

-The first adjustable seats were in the 1921 *Hudson Super Six*

The first popular song about a car was "In My Merry Oldsmobile." It was also the first song to be banned, when it was described what went on in the backseat.

In My Merry Oldsmobile

Young Johnny Steele has an Oldsmobile

He loves his dear little girl

She is the queen of his gas machine

She has his heart in a whirl

Now when they go for a spin, you know,

She tries to learn the auto, so

He lets her steer, while he gets her ear

And whispers soft and low...

They love to "spark" in the dark old park

As they go flying along

She says she knows why the motor goes

The "sparker" is awfully strong

Each day they "spoon" to the engine's tune

Their honeymoon will happen soon

He'll win Lucille with his Oldsmobile

And then he'll fondly croon...

Come away with me, Lucille,
In my merry Oldsmobile.

Down the road of life, we fly,
Automo-bubbling, you and I.

To the church we'll swiftly steal,
Then our wedding bells will peal,
You can go as far as you like with me
In my merry Oldsmobile.

Songwriters: Gus Edwards / Vincent Bryan   Oldsmobile lyrics © Warner/Chappell Music, Inc.

Chapter Forty-Three—U.S. Auto Production Figures 1897-1989

Production Figures for 1899-1901		Production Figures for 1901		Production Figures for 1902	
Columbia	1500	Locomobile	1500	Locomobile	2750
Locomobile	750	Winton	700	Oldsmobile	2500
Winton	100	Oldsmobile	425	Rambler	1500
Packard	49	White	193	White	385
Stanley automobile	30	Autocar	140	Knox	250
Stearns	20	Knox	100	Packard	179
Knox	15	Packard	81	Stanley automobile	170
Oldsmobile	11	Stanley automobile	80	Union	60

Production Figures for 1903		Production Figures for 1904		Production Figures for 1905	
Oldsmobile	4000	Oldsmobile	5508	Oldsmobile	6500

Cadillac	2497	Cadillac	2457	Cadillac	3942
Ford	1708	Rambler	2342	Rambler	3807
Pope-Hartford	1500	Ford	1685	Ford	1599
Rambler	1350	White	710	Franklin	1098
Winton	850	Stanley automobile	550	White	1015
White	502	Franklin	400	REO Motor Car Company	864
Knox	500	Packard	250	Maxwell	823

Production Figures for 1906		**Production Figures for 1907**		**Production Figures for 1908**	
Ford	8729	Ford	14,887	Ford	10,202
Cadillac	3559	Buick	4641	Buick	8820
Rambler	2765	REO Motor Car Company	3967	Studebaker	8132
REO Motor Car Company	2458	Maxwell	3785	Maxwell	4455

Maxwell	2161	Rambler	3201	REO Motor Car Company	4105
Oldsmobile	1600	Cadillac	2884	Rambler	3597
White	1534	Franklin	1509	Cadillac	2377
Buick	1400	Packard	1403	Franklin	1895

Production Figures for 1909		**Production Figures for 1910**		**Production Figures for 1911**	
Ford	17,771	Ford	32,053	Ford	69,762
Buick	14,606	Buick	30,525	Studebaker/EMF	26,827
Maxwell	9460	Willys-Overland	15,598	Willys-Overland	18,745
Studebaker/EMF	7960	Studebaker/EMF	15,020	Maxwell	16,000
Cadillac	7868	Cadillac	10,039	Buick	13,389
REO Motor Car Company	6592	Maxwell	10,000	Cadillac	10,071
Oldsmobile	6575	Brush	10,000	Hudson	6486

Willys-Overland	4907	REO Motor Car Company	6588	Chalmers	6250

Production Figures for 1912		**Production Figures for 1913**		**Production Figures for 1914**	
Ford	78,440	Ford	168,220	Ford	308,162
Willys-Overland	28,572	Willys-Overland	37,422	Willys-Overland	48,461
Studebaker/EMF	28,032	Studebaker	31,994	Studebaker	35,374
Buick	19,812	Buick	26,666	Buick	32,889
Cadillac	12,708	Cadillac	17,284	Maxwell	18,000
Hupmobile	7640	Maxwell	17,000	REO Motor Car Company	13,516
REO Motor Car Company	6342	Hupmobile	12,543	Jeffery	10,417
Oakland	5838	REO Motor Car Company	7647	Hupmobile	10,318

Production Figures for 1915		**Production Figures for 1916**		**Production Figures for 1917**	

Ford	501,492	Ford	734,811	Ford	622,351
Willys-Overland	91,904	Willys-Overland	140,111	Willys-Overland	130,988
Dodge	45,000	Buick	124,834	Buick	115,267
Maxwell	44,000	Dodge	71,400	Chevrolet	111,877
Buick	43,946	Chevrolet	70,701	Dodge	90,000
Studebaker	41,243	Maxwell	69,000	Maxwell	75,000
Cadillac	20,404	Studebaker	65,536	Studebaker	39,686
Saxon	19,000	Saxon	27,800	Oakland	33,171

Production Figures for 1918   **Production Figures for 1919**   **Production Figures for 1920**

Ford	435,898	Ford	820,445	Ford	806,040
Willys-Overland	88,753	Chevrolet	129,118	Chevrolet	146,243
Chevrolet	88,717	Buick	119,310	Dodge	141,000
Buick	77,691	Dodge	106,000	Buick	115,176

Dodge	62,000	Willys-Overland	80,853	Willys-Overland	105,025
Maxwell	34,000	Oakland	52,124	Studebaker	48,831
Oakland	27,757	Maxwell	50,000	Hudson/Essex	45,937
Oldsmobile	19,169	Oldsmobile	39,042	Chandler	45,000

Production Figures for 1921		**Production Figures for 1922**		**Production Figures for 1923**	
Ford	1,275,618	Ford	1,147,028	Ford	1,831,128
Chevrolet	130,855	Dodge	152,653	Chevrolet	323,182
Buick	82,930	Chevrolet	138,932	Buick	201,572
Dodge	81,000	Buick	123,152	Willys-Overland	196,038
Studebaker	65,023	Studebaker	105,005	Durant	172,000
Willys-Overland	48,016	Willys-Overland	95,410	Dodge	151,000
Hudson/Essex	27,143	Durant	55,300	Studebaker	146,238

Nash	20,850	Maxwell/Chalmers	44,811	Hudson/Essex	88,914

Production Figures for 1924 | **Production Figures for 1925** | **Production Figures for 1926**

Production Figures for 1924		Production Figures for 1925		Production Figures for 1926	
Ford	1,720,795	Ford	1,669,847	Ford	1,426,612
Chevrolet	264,868	Chevrolet	306,479	Chevrolet	547,724
Dodge	193,861	Hudson/Essex	269,474	Buick	266,753
Willys-Overland	163,000	Willys-Overland	215,000	Dodge	265,000
Buick	160,411	Dodge	201,000	Hudson/Essex	227,508
Hudson/Essex	133,950	Buick	192,100	Willys-Overland/Whippet	182,000
Durant	111,000	Studebaker	133,104	Chrysler	135,520
Studebaker	105,387	Chrysler/Maxwell	132,343	Pontiac/Oakland	133,604

Production Figures for 1927 | **Production Figures for 1928** | **Production Figures for 1929**

Chevrolet	1,001,820	Chevrolet	1,193,212	Ford	1,507,132
Ford	367,213	Ford	607,592	Chevrolet	1,328,605
Hudson/Essex	276,414	Willys-Overland/Whippet	315,000	Hudson/Essex	300,962
Buick	255,160	Hudson/Essex	282,203	Willys-Overland/Whippet	242,000
Pontiac/Oakland	188,168	Pontiac/Oakland	244,584	Pontiac/Oakland	211,054
Willys-Overland/Whippet	188,000	Buick	221,758	Buick	196,104
Chrysler	182,195	Chrysler	160,670	Dodge	124,557
Dodge	180,000	Nash	138,137	Nash	116,622

Production Figures for 1930		**Production Figures for 1931**		**Production Figures for 1932**	
Ford	1,140,710	Chevrolet	619,554	Chevrolet	313,404
Chevrolet	640,980	Ford	615,455	Ford	210,824
Buick	181,743	Buick	138,965	Plymouth	186,106

Studebaker	123,215	Studebaker	96,173	Hudson/ Essex	57,550
Hudson/Essex	113,898	Pontiac	84,708	Buick	56,790
Plymouth	108,350	Plymouth	75,510	Pontiac	45,340
Dodge	90,755	Willys	65,800	Nash	30,834
Chrysler	77,881	Chrysler	65,500	Willys	27,800
Production Figures for 1933		**Production Figures for 1934**		**Production Figures for 1935**	
Chevrolet	486,261	Ford	563,921	Ford	820,253
Ford	334,969	Chevrolet	551,191	Chevrolet	548,215
Plymouth	298,557	Plymouth	321,171	Plymouth	350,884
Dodge	106,103	Dodge	95,011	Pontiac	178,770
Pontiac	90,198	Hudson/ Terraplane	85,835	Dodge	158,999
Buick	46,924	Oldsmobile	79,814	Oldsmobile	126,768
Studebaker/	43,024	Pontiac	78,859	Hudson/	101,080

Rockne				Terraplane	
Hudson/Essex	40,982	Buick	71,009	Buick	53,249
Production Figures for 1936		**Production Figures for 1937**		**Production Figures for 1938**	
Ford	930,778	Ford	942,005	Chevrolet	465,158
Chevrolet	918,278	Chevrolet	815,375	Ford	410,263
Plymouth	520,025	Plymouth	566,128	Plymouth	285,704
Dodge	263,647	Dodge	295,047	Buick	168,689
Oldsmobile	200,546	Pontiac	236,189	Dodge	114,529
Pontiac	176,270	Buick	220,346	Oldsmobile	99,951
Buick	168,596	Oldsmobile	200,886	Pontiac	97,139
Hudson/Terraplane	123,266	Packard	122,593	Packard	55,718
Production Figures for 1939		**Production Figures for 1940**		**Production Figures for 1941**	
Chevrolet	577,278	Chevrolet	764,616	Chevrolet	1,008,976

Ford	487,031	Ford	541,896	Ford	691,455
Plymouth	423,850	Plymouth	430,208	Plymouth	522,080
Buick	208,259	Buick	278,784	Buick	374,196
Dodge	186,474	Dodge	225,595	Pontiac	330,061
Pontiac	144,340	Pontiac	217,001	Oldsmobile	270,040
Oldsmobile	137,249	Oldsmobile	192,692	Dodge	215,575
Studebaker	85,834	Studebaker	107,185	Chrysler	161,704

Production Figures for 1942 / **Production Figures for 1946** / **Production Figures for 1947**

Chevrolet	254,885	Ford	468,022	Chevrolet	671,546
Ford	160,432	Chevrolet	398,028	Ford	429,674
Plymouth	152,427	Plymouth	264,660	Plymouth	382,290
Buick	92,573	Dodge	163,490	Buick	272,827
Pontiac	83,555	Buick	153,627	Dodge	243,160
Dodge	68,522	Pontiac	137,640	Pontiac	230,600

Oldsmobile	67,783	Oldsmobile	117,623	Oldsmobile	193,895
Studebaker	50,678	Nash	94,000	Studebaker	161,496
Hudson	40,661	Hudson	91,039	Chrysler	119,260
Chrysler	36,586	Mercury	86,608	Nash	101,000
Packard	33,776	Chrysler	83,310	Hudson	92,038
Nash	31,780	DeSoto	66,900	DeSoto	87,000
DeSoto	24,015	Packard	30,793	Mercury	85,383
Mercury	22,816	Cadillac	29,214	Kaiser	70,474
Cadillac	16,511	Studebaker	19,275	Frazer	68,775
Lincoln	6,547	Lincoln	16,645	Cadillac	61,926
Crosley	1,029	Crosley	4999	Packard	51,086
				Lincoln	21,460
				Crosley	19,344
Production Figures for 1948		**Production Figures for 1949**		**Production Figures for 1950**	

Chevrolet	696,449	Ford	1,118,308	Chevrolet	1,498,590
Ford	430,198	Chevrolet	1,010,013	Ford	1,208,912
Plymouth	412,540	Plymouth	520,385	Plymouth	610,954
Dodge	243,340	Buick	409,138	Buick	588,439
Pontiac	235,419	Pontiac	304,819	Pontiac	446,429
Buick	213,599	Mercury	301,319	Oldsmobile	408,060
Studebaker	184,993	Oldsmobile	288,310	Dodge	341,797
Oldsmobile	172,852	Dodge	256,857	Studebaker	320,884
Chrysler	130,110	Hudson	159,100	Mercury	293,658
Hudson	117,200	Nash	135,328	Chrysler	179,299;
Nash	110,000	Studebaker	129,301	Nash	171,782
DeSoto	98,890	Chrysler	124,218	DeSoto	136,203
Packard	92,251	Packard	116,955	Hudson	121,408
Kaiser	91,851	DeSoto	95,051	Cadillac	103,857

Cadillac	52,706	Cadillac	92,554	Packard	42,627
Mercury	50,268	Kaiser	79,947	Lincoln	28,190
Frazer	48,071	Lincoln	73,507	Kaiser	15,228
Crosley	26,239	Frazer	21,223	Crosley	6,792
Lincoln	7,769	Crosley	7,431	Frazer	3,700

Production Figures for 1951		**Production Figures for 1952**		**Production Figures for 1953**	
Chevrolet	1,229,986	Chevrolet	818,142	Chevrolet	1,346,475
Ford	1,013,381	Ford	671,733	Ford	1,247,542
Plymouth	611,000	Plymouth	396,000	Plymouth	650,451
Buick	404,657	Buick	303,745	Buick	488,755
Pontiac	370,159	Pontiac	271,373	Pontiac	418,619
Mercury	310,387	Oldsmobile	213,490	Oldsmobile	334,462
Dodge	290,000	Dodge	206,000	Dodge	320,008
Oldsmobile	285,615	Mercury	172,087	Mercury	305,863

Studebaker	246,195	Studebaker	167,662	Chrysler	170,006
Nash	205,307	Nash	154,291	Studebaker	151,576
Chrysler	163,613	Cadillac	90,259	DeSoto	132,104
Kaiser	139,452	DeSoto	88,000	Nash	121,793
Hudson	131,915	Chrysler	87,470	Cadillac	109,651
Cadillac	110,340	Hudson	70,000	Packard	90,252
DeSoto	106,000	Packard	62,921	Hudson	66,143
Packard	100,713	Kaiser	32,131	Willys	42,224
Henry J	81,942	Willys	31,363	Lincoln	40,762
Lincoln	32,574	Henry J	30,585	Kaiser	27,652
Frazer	10,214	Lincoln	27,271	Henry J	16,672
Crosley	6,614	Crosley	6,614	Metropolitan	743 (calendar year sales)

Production Figures for 1954	**Production Figures for 1955**	**Production Figures for 1956**

Ford	1,165,942	Chevrolet	1,704,667	Chevrolet	1,567,117
Chevrolet	1,143,561	Ford	1,451,157	Ford	1,408,478
Plymouth	463,148	Buick	738,814	Buick	572,024
Buick	444,609	Plymouth	705,455	Plymouth	571,634
Oldsmobile	354,001	Oldsmobile	583,179	Oldsmobile	485,458
Pontiac	287,744	Pontiac	554,090	Pontiac	405,730
Mercury	259,305	Mercury	329,808	Mercury	327,943
Dodge	154,648	Dodge	276,936	Dodge	240,686
Chrysler	105,030	Chrysler	152,777	Cadillac	154,577
Cadillac	96,680	Cadillac	140,777	Chrysler	128,322
Nash	91,121	Studebaker	116,333	DeSoto	109,442
DeSoto	76,580	DeSoto	115,485	Nash	83,420
Studebaker	68,708	Nash	96,156	Studebaker	69,593
Hudson	50,660	Packard	55,247	Lincoln	50,322

Lincoln	36,993	Hudson	45,535	Hudson	22,588
Packard	31,291	Lincoln	27,222	Clipper	18,482
Metropolitan	13,162	Imperial	11,432	Imperial	10,684
Willys	11,856	Willys	6,565	Packard	10,353
Kaiser	8,539	Metropolitan	6,096	Metropolitan	9,068
Henry J	1,123	Kaiser	1,291	Continental	2,550

Production Figures for 1957		**Production Figures for 1958**		**Production Figures for 1959**	
Ford	1,676,449	Chevrolet	1,142,460	Chevrolet	1,462,140
Chevrolet	1,505,910	Ford	987,945	Ford	1,450,953
Plymouth	726,009	Plymouth	443,799	Plymouth	458,261
Buick	405,086	Oldsmobile	294,374	Pontiac	383,320
Oldsmobile	384,390	Buick	241,892	Oldsmobile	382,865
Pontiac	334,041	Pontiac	217,303	Rambler	374,240
Dodge	287,608	Rambler	162,182	Buick	285,089

Mercury	286,163	Dodge	137,861	Dodge	156,385
Cadillac	146,841	Mercury	133,271	Mercury	150,000
DeSoto	126,514	Cadillac	121,778	Cadillac	142,272
Chrysler	122,273	Chrysler	63,681	Studebaker	126,156
Rambler	91,469	Edsel	63,110	Chrysler	69,970
Studebaker	63,101	DeSoto	49,445	DeSoto	45,734
Lincoln	41,123	Studebaker	44,759	Edsel	44,891
Imperial	37,593	Lincoln	17,134	Lincoln/ Continental	26,906
Metropolitan	15,317	Imperial	16,133	Metropolitan	22,209
Nash	10,330	Metropolitan	13,128	Imperial	17,269
Packard	4,809	Continental	12,550	Checker	1,050
Hudson	4,180	Packard	2,622		
Continental	462				
Production Figures for		**Production Figures for**		**Production Figures for**	

1960		**1961**		**1962**	
Chevrolet	1,653,168	Ford	1,338,790	Chevrolet	2,061,677
Ford	1,439,370	Chevrolet	1,318,014	Ford	1,476,031
Plymouth	483,969	Rambler	377,902	Pontiac	521,933
Rambler	458,841	Plymouth	356,257	Rambler	442,346
Pontiac	396,716	Pontiac	340,635	Oldsmobile	428,853
Dodge	367,804	Oldsmobile	317,548	Buick	399,526
Oldsmobile	347,142	Mercury	317,351	Mercury	341,366
Mercury	271,331	Buick	276,754	Plymouth	339,527
Buick	253,807	Dodge	269,367	Dodge	240,484
Cadillac	142,184	Cadillac	138,379	Cadillac	160,840
Studebaker	120,465	Chrysler	96,454	Chrysler	128,921
Chrysler	77,285	Studebaker	59,713	Studebaker	89,318
DeSoto	26,081	Lincoln	25,164	Lincoln	31,061
Lincoln/Con	24,820	Imperial	12,258	Imperial	14,337

tinental					
Imperial	17,719	DeSoto	3,034	Checker	1,230
Metropolitan	13,103	Metropolitan	969	Metropolitan	420
Edsel	3,008	Checker	860		

Production Figures for 1963		**Production Figures for 1964**		**Production Figures for 1965**	
Chevrolet	2,237,201	Chevrolet	2,318,619	Chevrolet	2,375,118
Ford	1,525,404	Ford	1,594,053	Ford	2,170,795
Pontiac	590,071	Pontiac	715,261	Pontiac	802,000
Plymouth	488,448	Plymouth	551,633	Plymouth	728,228
Oldsmobile	476,753	Buick	510,490	Buick	600,145
Rambler	464,126	Dodge	501,781	Oldsmobile	591,701
Buick	457,818	Oldsmobile	493,991	Dodge	489,065
Dodge	446,129	Rambler	393,859	Rambler	391,366
Mercury	301,581	Mercury	298,609	Mercury	346,751

Cadillac	163,174	
Chrysler	128,937	
Studebaker	69,555	
Lincoln	31,233	
Imperial	14,121	
Checker	1,080	

Cadillac	165,909	
Chrysler	153,319	
Studebaker	36,697	
Lincoln	36,297	
Imperial	23,295	
Checker	960	

Chrysler	206,089	
Cadillac	182,435	
Lincoln	40,180	
Studebaker	19,435	
Imperial	18,409	
Checker	930	
Excalibur	56	
Avanti II	21	

Production Figures for 1966

Ford	2,212,415
Chevrolet	2,206,639
Pontiac	831,331
Plymouth	687,514
Dodge	632,658

Production Figures for 1967

Chevrolet	1,948,410
Ford	1,730,224
Pontiac	782,734
Plymouth	638,075
Buick	562,507

Production Figures for 1968

Chevrolet	2,139,290
Ford	1,753,334
Pontiac	910,977
Plymouth	790,239
Buick	651,823

Oldsmobile	578,385	Oldsmobile	548,390	Dodge	627,533
Buick	553,870	Dodge	465,732	Oldsmobile	562,459
Mercury	343,149	Mercury	354,923	Mercury	360,467
Rambler/ AMC	341,951	Rambler/ AMC	302,945	AMC/ Rambler	446,781
Chrysler	264,848	Chrysler	218,742	Chrysler	264,853
Cadillac	196,685	Cadillac	200,000	Cadillac	230,003
Lincoln	54,755	Lincoln	45,667	Lincoln	46,904
Imperial	13,742	Imperial	17,620	Imperial	15,367
Studebaker	8,947	Shelby	3,225	Shelby	4,451
Shelby	2,378	Checker	950	Checker	992
Checker	1,056	Excalibur	71	Avanti II	89
Avanti II	98	Avanti II	60	Excalibur	57
Excalibur	90				
Production Figures for		**Production Figures for**		**Production Figures for**	

1969		1970		1971	
Chevrolet	2,092,947	Ford	2,096,184	Ford	2,054,351
Ford	1,826,777	Chevrolet	1,451,305	Chevrolet	1,830,319
Pontiac	870,081	Plymouth	747,508	Plymouth	702,113
Plymouth	751,134	Pontiac	690,953	Pontiac	586,856
Buick	665,422	Buick	666,501	Oldsmobile	567,891
Oldsmobile	635,241	Oldsmobile	633,981	Dodge	551,386
Dodge	611,645	Dodge	543,019	Buick	551,188
Mercury	398,262	Mercury	324,716	Mercury	365,310
AMC/ Rambler	309,000	AMC	276,000	AMC	244,758
Chrysler	260,773	Cadillac	238,744	Cadillac	188,537
Cadillac	223,237	Chrysler	180,777	Chrysler	175,118
Lincoln	61,378	Lincoln	59,127	Lincoln	62,642
Imperial	22,103	Imperial	11,822	Imperial	11,558

Shelby	3,150	Checker	397	Checker	500
Checker	760	Avanti II	111	Avanti II	107
Avanti II	103	Excalibur	37		
Excalibur	91				

Production Figures for 1972		**Production Figures for 1973**		**Production Figures for 1974**	
Chevrolet	2,420,564	Chevrolet	2,579,509	Chevrolet	2,333,839
Ford	2,246,563	Ford	2,349,815	Ford	2,179,791
Oldsmobile	762,199	Oldsmobile	922,771	Plymouth	739,894
Plymouth	756,605	Pontiac	919,870	Oldsmobile	581,195
Pontiac	706,978	Plymouth	882,196	Pontiac	580,045
Buick	679,921	Buick	821,165	Buick	495,063
Dodge	577,870	Dodge	665,536	Dodge	477,728
Mercury	441,964	Mercury	486,470	AMC	431,798
Cadillac	267,787	AMC	392,105	Mercury	403,977

AMC	258,134	Cadillac	304,839	Cadillac	242,330
Chrysler	204,704	Chrysler	234,223	Chrysler	117,373
Lincoln	94,560	Lincoln	128,073	Lincoln	93,983
Imperial	15,804	Imperial	16,729	Imperial	14,426
Checker	850	Checker	900	Checker	900
Avanti II	127	Excalibur	122	Avanti II	123
Excalibur	65	Avanti II	106	Excalibur	118

Production Figures for 1975		**Production Figures for 1976**		**Production Figures for 1977**	
Chevrolet	1,755,773	Chevrolet	2,103,862	Chevrolet	2,543,153
Ford	1,569,608	Ford	1,861,537	Ford	1,840,427
Oldsmobile	631,795	Oldsmobile	891,368	Oldsmobile	1,135,803
Pontiac	531,922	Pontiac	746,430	Pontiac	850,620
Buick	481,768	Buick	737,466	Buick	845,234
Plymouth	454,105	Plymouth	519,962	Plymouth	546,132

Mercury	404,650	Mercury	480,361	Dodge	526,254
Dodge	377,462	Dodge	430,641	Mercury	521,909
Cadillac	264,732	Cadillac	309,139	Chrysler	399,297
Chrysler	251,549	AMC	283,577	Cadillac	358,488
AMC	241,501	Chrysler	222,153	Lincoln	191,355
Lincoln	101,843	Lincoln	124,756	AMC	182,005
Imperial	8,830	Excalibur	184	Excalibur	237
Checker	450	Avanti II	156	Avanti II	146
Avanti II	125				
Excalibur	90				

Production Figures for 1978		**Production Figures for 1979**		**Production Figures for 1980**	
Chevrolet	2,375,436	Chevrolet	2,284,749	Chevrolet	2,288,745
Ford	1,923,655	Ford	1,835,937	Ford	1,162,275
Oldsmobile	1,015,805	Oldsmobile	1,068,154	Oldsmobile	910,306

Production Figures for 1981	
Pontiac	900,380
Buick	803,187
Mercury	635,051
Plymouth	501,129
Dodge	467,720
Chrysler	354,029
Cadillac	349,684
Lincoln	169,620
AMC	137,860
Excalibur	263
Avanti II	165
Chevrolet	1,673,093
Ford	950,301

Production Figures for 1982	
Pontiac	907,434
Buick	727,275
Mercury	669,138
Dodge	404,266
Cadillac	383,138
Plymouth	372,449
Chrysler	349,450
Lincoln	189,546
AMC	169,439
Excalibur	367
Avanti II	142
Chevrolet	1,297,357
Ford	748,732

Production Figures for 1983	
Buick	854,011
Pontiac	770,100
Mercury	347,711
Dodge	308,638
Plymouth	290,974
Cadillac	230,028
AMC	199,613
Chrysler	164,510
Lincoln	74,908
Avanti II	168
Excalibur	93
Chevrolet	1,175,200
Oldsmobile	916,583

Oldsmobile	873,678	Buick	739,984	Buick	808,416
Buick	856,996	Oldsmobile	702,340	Ford	783,225
Pontiac	489,436	Pontiac	541,061	Mercury	359,594
Plymouth	393,633	Mercury	328,597	Pontiac	318,478
Mercury	375,756	Plymouth	247,936	Dodge	304,464
Dodge	340,899	Dodge	241,359	Cadillac	292,814
Cadillac	240,189	Cadillac	235,584	Plymouth	273,489
AMC	137,125	Chrysler	103,310	AMC	168,726
Lincoln	69,537	Lincoln	85,313	Chrysler	159,882
Chrysler	56,726	AMC	70,898	Lincoln	101,068
Excalibur	235	Excalibur	212	Avanti	100
Avanti II	200	Avanti II	200		
Production Figures for 1984		**Production Figures for 1985**		**Production Figures for 1986**	
Chevrolet	1,655,151	Chevrolet	1,418,098	Chevrolet	1,368,837

Ford	1,180,708	Oldsmobile	1,165,649	Ford	1,253,525
Oldsmobile	1,144,225	Ford	1,149,427	Oldsmobile	1,050,832
Buick	987,980	Buick	1,002,906	Buick	850,103
Pontiac	594,821	Pontiac	519,390	Pontiac	799,461
Mercury	475,381	Dodge	500,835	Dodge	450,365
Dodge	442,527	Chrysler	420,780	Mercury	399,240
Chrysler	375,853	Mercury	419,869	Chrysler	367,898
Plymouth	357,764	Plymouth	393,711	Plymouth	350,573
Cadillac	300,300	Cadillac	384,840	Cadillac	281,683
AMC	208,624	Lincoln	166,486	Lincoln	156,839
Lincoln	157,434	AMC	150,189	AMC	64,873
Avanti	287				

Production Figures for 1987		**Production Figures for 1988**		**Production Figures for 1989**	
Chevrolet	1,384,214	Ford	1,331,489	Chevrolet	1,275,498

Ford	1,176,775	Chevrolet	1,236,316	Ford	1,234,954
Pontiac	724,289	Pontiac	680,714	Pontiac	801,600
Oldsmobile	670,880	Oldsmobile	535,015	Oldsmobile	533,818
Buick	648,689	Dodge	489,645	Buick	506,787
Dodge	501,926	Buick	458,768	Dodge	481,139
Plymouth	443,806	Plymouth	336,070	Mercury	294,899
Chrysler	360,613	Mercury	298,859	Cadillac	276,330
Mercury	315,147	Lincoln	280,659	Plymouth	268,442
Cadillac	282,582	Chrysler	278,287	Chrysler	224,097
Lincoln	109,366	Cadillac	270,844	Lincoln	215,966
AMC	36,336	Avanti	150	Avanti	350

Afterword

The author, Alan Naldrett, is still in search of someone who will donate to him a Duesenberg Model J. Thank you!